CW01511922

Welsh
Folk Tales
of Coast
& Sea

Welsh
Folk Tales
of Coast
& Sea

PETER
STEVENSON

The
History
Press

To Ailsa Mair, the Mesolithic Mermaid

First published 2025

Reprinted 2025

The History Press
97 St George's Place, Cheltenham,
Gloucestershire, GL50 3QB
www.thehistorypress.co.uk

© Peter Stevenson, 2025

The right of Peter Stevenson to be identified as the Author
of this work has been asserted in accordance with the
Copyright, Designs and Patents Act 1988.

All rights reserved. No part of this book may be reprinted
or reproduced or utilised in any form or by any electronic,
mechanical or other means, now known or hereafter invented,
including photocopying and recording, or in any information
storage or retrieval system, without the permission in writing
from the Publishers.

British Library Cataloguing in Publication Data.
A catalogue record for this book is available from the British Library.

ISBN 978 1 80399 662 2

Typesetting and origination by The History Press
Printed and bound in Great Britain by TJ Books, Padstow, Cornwall.

Trees for LYfe

Contents

Taith Môrwen

Môrwen's Travels

Chwedl Dŵr/The Fairy Tale of Water

Ar lan y môr mae rhosys cochion,
Ar lan y môr mae lilis gwynion,
Ar lan y môr mae 'nghariad inne,
Yn cysgu'r nos a chodi'r bore.

On the seashore are red roses,
On the seashore are white lilies,
On the seashore is my love,
Sleeping at night and rising in
the morning.

Traditional

Y Môr/The Sea

Folk tales wash up with the tide along the
Welsh coast like streams of thought, waiting for
passing beachcombers to spot them amongst the
seaweed, crab legs and plastic bottles. They
carry memories of flood myths, submerged
forests, a Welsh utopia, sea monsters,
water horses, fairy islands, love
potions, drunken mermaids,
starling murmurations,

rockpool worlds, Welsh-speaking dolphins, healing wells, coastal schooners, sewage discharges and the sweet sadness of seaside towns in winter. Folk tales are looking glasses into the lives of those who hunted, foraged and lived in the forests and marshes after the ice melted and flooded the land.

Mesolithic people lived as nomads along the Welsh coast between 4,000 and 12,000 years ago. They navigated by the stars as they followed the seasons, and left flint tools, charcoal deposits, cores, flakes and limpet middens as an archaeological patchwork quilt of their ephemeral lives.

There are folk tales of lost lands in Cardigan Bay, Conwy estuary and Cynffig dunes, and submerged forests off Abergele, Amroth, Borth, Goldcliff, Llanrhystud, Lydstep, Newgale, Rhyl and Whitesands. Tree stumps on the beach at Ynyslas have been dated to 5,500 years ago, while those a kilometre away at Borth are around 4,000 years old, when the first indications of farming appeared in Wales. The inter-tidal strip at Whitesands reveals evidence of auroch, red deer and brown bear, while pig bones have been found at Lydstep, and a much older woolly mammoth's jawbone at Holyhead.

Cartographers have tried to make order out of the chaos of mythology and memory, yet the history of early map-making contains ghost islands. A mid-thirteenth-century map of the British Isles given to the Bodleian Library by Richard Gough, who bought it from the antiquarian 'Honest Tom' Martin in 1774, shows two large islands off the Ceredigion and Meirionnydd coast. The romantic narrative suggests they may be a memory of the submerged land of Cantre'r Gwaelod (Chapter 3), or the green fairy islands that disappear when mariners sail towards them (Chapters 1, 19 and 20). However, the map doesn't show Pen Llŷn, and if you stand on Aberystwyth prom and stare at the horizon, the peninsula looks suspiciously like islands.

Lewis Morris, antiquary, literary scholar and self-taught hydrographer from Ynys Môn, produced a series of more accurate maps of the coastal seas between Llandudno and Milford Haven in the mid-1700s in order to make navigation safer (Chapter 9).

Archaeologists and geomythologists have now mapped the layers of land beneath the Welsh seas to help understand the geography of river valleys and forests in the Mesolithic age.

As nomadic lifestyles gave way to farming, Iron Age hillforts appeared along the west coast. Romans built ports at Caerleon and Caernarfon, Normans settled along the Glamorgan coast, and Vikings left their names on the Western islands. People fished with seine nets, lave nets, lobster creels, crab pots and coracles, and traded using smacks, ketches and sloops. Water connected people in a land with poor roads and expensive tolls.

When Michael Faraday visited Wales in the early 1800s, he revealed the extent of industrial pollution caused by the copper industry in Swansea and showed how lead waste from mines in the mid-Wales hills flowed downriver and poisoned the sea. These issues may not be new, but the scale and pace of twenty-first-century damage is. Recently, Dŵr Cymru admitted pouring untreated sewage into Welsh rivers, while Merseyside Docks dumped spoil in a Welsh Marine Conservation Area. The sea is seen as a resource to be exploited for leisure, mining, industrial fishing and renewable energy, whereas *she* is asserting her natural, moral and legal rights. She is writing her own Mabinogi.

In this book, the folk tales of coast and sea are seen through the eyes of a 7,000-year-old Marsh Girl who will step through the veil between her Mesolithic world and ours on Calan Gaeaf, the first of November. This is not time travel. It happens in Wales on this one day of the year, when we tell stories of those who walked the land as spirits and ghosts before us.

Marsh Girl has a name: Môrwen. White Sea. She has been raised on stories that contained fragments of the imaginations of the people who told them, like tool-marks on a microlith, a footprint in the peat or a carving on a tree. Flood myths are real to Môrwen. They happened in her lifetime.

Just ask those mountains over there.

They have heard these stories before.

They have witnessed the floods.

1

The Mesolithic Mermaid and the Welsh Utopia

BAE CEREDIGION/CARDIGAN BAY

The Welsh Utopia

Amser maith yn ôl/ A long time ago.

The shallow sea in Cardigan Bay, from Pen Llŷn in the north to Ceredigion in the west, was once a mix of forests, lakes, rivers, swamps and saltmarsh. The nomadic people who lived there cared for their land, yet never thought they owned it. They foraged and hunted, treated animals as equals and left offerings in exchange for anything they took. This was the land of Plant Rhys Ddwfn, the Children of Rhys the Deep. Not deep below the sea – Rhys was a thinker, a dreamer, a philosopher.

He reasoned that if the mean giants who lived in the mountains ever saw his land, they would destroy it. So, he devised a cunning plan. He planted a hedge of herbs along what is now the west Welsh coast, to hide his land from their prying eyes. Only if the giants stood on the one small clump of this herb that grew away from the coast would they see Rhys' world, but as they had no idea where this piece of turf was, all they saw was rain.

Rhys' children cared for each other, their numbers grew, food became scarce, and the giants heard the distant rumble of empty bellies, although they mistook it for the anger of the gods.

They turned to crafts, became toolmakers, wood carvers and basket makers, and travelled by sea to the markets in Ceredigion to trade their goods – but as soon as they were seen, prices went up. They traded with Gruffydd ap Einion, a radical free-thinker who dreamed of a fairer world. After many years, they took him to the clump of herbs, where he saw Rhys' land with all the knowledge and wisdom in the world archived safely in forests and books. Preachers and politicians were few. Choughs and kestrels hung in the air. The land was rich beyond dreams, the utopia he had long dreamed of.

Gruffydd asked how they kept themselves safe from crime, and they explained that Rhys' herbs hid them from an angry creature with horns, snakes and a sword that spewed toxic venom at anything it disagreed with.

When Gruffydd stepped away from the herbs, he lost sight of Rhys' land, though he never forgot there was a better world out there in Cardigan Bay. Rhys' children traded with their friend all his life, until one day they came to the market to find Gruffydd's hair had turned to snow and he had passed over to the Otherworld.

As the floodwaters lapped at their feet, Rhys' children turned to a nomadic life, following the seasons and tracks of animals along the water's edge, fishing and foraging, carrying the bones of their ancestors to remember their stories. For pity the people who have forgotten their myths.

Marsh Girl

Seven thousand years ago, Môrwen was born into this world of rising floodwaters as one of Rhys' children. She fishes for salmon, forages for hazelnuts, scavenges for honey, digs for celandine tubers, carves wooden spoons from holly, weaves baskets from rushes, draws animals on stone with charcoal and makes pigments from crushed rock. She knows the movements of the deer herd and follows their tracks and scents through the forest. She collects antlers shed by the old stags and sharpens the points into axe-heads with flint tools. She has no need to keep deer in enclosures – they come when she calls. She is sharp as flint, moulded from the dust of time. She has spent so much of her life up to her waist in water, friends call her Marsh Girl.

In the evenings Môrwen huddles at Nan's feet to hear stories of the mean giants of the mountains, the mischievous old women who make potions from the herbs of the forest, the green man of no one's land, and the girls who transform into fish, birds and wolves. She draws mammoths chased by little stick men who glory in the spilling of blood and praise themselves in poetry and song. She has never seen a mammoth, but Nan's words paint them in her imagination. They become real when she sketches on rocks, until the rain and floods wash them away.

One day Môrwen is foraging along the riverbank when a storm gathers out to sea, waves crash over the beach, and the forest fills with floodwater. She runs for high ground but slips into the swamp and sinks up to her shoulders. She grabs hold of a clump of tough reeds and bends her knees to stop herself being sucked further into the wet peat. She hangs there for what seems like hours until the wind abates and she hauls herself on to the dry forest floor, where she lies breathless, staring at the stars.

Thoughts swish round her mind like waves. Is the sea having a laugh? Why are Rhys' children losing their way of life to the floods? Will the sun set fire to the trees? Do her descendants have answers in the future? She knows there is a future, for she has stepped through the veil on Calan Gaeaf before.

The Cardigan Bay Mermaids

*Amser maith yn ôl/*A long time ago.

In the Old Welsh Dreamtime, when people were people and fish were fish, three brothers lived in a yellow stone farmhouse overlooking the forests and swamps of Cardigan Bay. Eldest Brother ploughed the land and had honey on his bread, Middle Brother farmed the sea and had salt in his porridge, whilst Little Brother wandered the old Welsh Tramping Road with a tune on his lips, head in the clouds and feet in the mud, and when his belly rumbled he asked his brothers for food. The two hard-working brothers grumbled.

'Little Brother?' said Eldest Brother, holding a small pig on a rope, 'Take this enchanted pig, sell it for money – and don't swap it for anything that makes wishes come true! You know what happens to wishes in fairy tales?'

Little Brother nodded and set off along the Old Welsh Tramping Road with a tune on his lips, an enchanted pig on a rope and no thoughts about wishes in fairy tales. He walked until he came to a deep dark wood, and in the middle of the wood found a crooked lime-washed house with a red door. In the doorway stood an old woman with a thousand wrinkles round her eyes and a single yellow tooth wobbling unnervingly in the breeze from her breath.

'Would you like to buy an enchanted pig?' asked Little Brother.

The woman pulled on the one grey hair in the middle of her chin, and drooled. 'Mmm, roast pork! I'll swap your pig for my enchanted handmill. It will make your wildest wishes come true.' The pig hid behind Little Brother's legs.

Little Brother completely forgot about wishes in fairy tales, and the deal was done. He set off back along the Old Welsh Tramping Road to the shores of Cardigan Bay with a tune on his lips and an enchanted handmill under his arm, and as the red door closed, he heard the squealing of a pig.

Little Brother returned to the yellow stone farmhouse overlooking Cardigan Bay, and thought, 'I'd like a cottage of my own.'

'Little mill, little mill, grind me a handsome house.

'Little mill, little mill, grind it without a mouse.'

And the handmill ground out a pink-washed longhouse with a table, a chair, a bottle of wine and a roaring fire. Now he would never need to ask his brothers for food again.

Eldest Brother looked out of the window of the yellow stone farmhouse and saw a pink-washed longhouse that wasn't there yesterday. He knocked on the door and there stood Little Brother.

'Little Brother, last time I saw you, you were poor as a church mouse, now you're rich as a lord. Where has all this money come from?' and Eldest Brother poured out a bottle of home-brewed beer. Before Little Brother passed out, he told Eldest Brother all about the handmill.

Eldest Brother took the handmill home to the yellow stone farmhouse, placed it on the kitchen table and made a wish.

'Little mill, little mill, grind me maids and ale.

'Little mill, little mill, grind them dark and pale.

Oh – and a little fish for my tea.'

Eldest Brother was a simple man.

The handmill began to grind out strong beer till it covered the floor, then a dark girl with the tail of a fish, followed by a pale girl, also with a fish tail. Soon the beer covered Eldest Brother's feet, his knees, waist, belly, chin, and mermaids were frolicking in the sea of ale, so he shouted '*Stop!*', but the handmill continued grinding 'til the door burst open and a river of beer and mermaids flowed down into Cardigan Bay, flooding the forests and swamps until Eldest Brother drowned, as he would have wanted to go, with a drunken fish-girl on either arm.

Little Brother awoke with a headache to the sound of rowdy mermaids frolicking in the floodwater. At that moment, Middle Brother, the one who ploughed the sea, sailed into Cardigan Bay in his red-masted ship with a cargo of salt from a faraway land, to find a sea full of mermaids singing rude sea shanties and impolitely inviting him to remove his trousers and join them.

He dropped anchor, waded ashore and went to the yellow stone farmhouse, where he found Little Brother holding a handmill that was still grinding out beer and mermaids.

'Little Brother, when I left, this was all land and now it's water. And where have all these drunken mermaids come from?' Middle Brother produced a bottle of smuggled Jamaican Rum and before he passed out, Little Brother told Middle Brother about the handmill.

Middle Brother took the handmill back to his ship, placed it on the deck and made a wish.

'Little mill, little mill, grind me salty salt.

'Little mill, little mill, grind it without a halt!' for with an endless supply of salt, he would never have to sail to faraway lands, live on mouldy biscuits or be looted by pirates.

The handmill began to grind until the deck was covered in salt, and soon it covered his feet, knees, waist, belly and chin, so he climbed the red mast but the salt climbed higher and under its weight, the ship sank to the bottom of the sea, where Middle Brother, like Eldest Brother, drowned in the arms of rowdy mermaids.

Little Brother woke with another headache and went to the seashore for a drink, but the water tasted of salt, beer and mermaids, and now he remembered. 'Is *this* what happens to wishes in fairy tales?'

So he set off along the old Welsh Tramping Road with a tune on his lips, head in the clouds and feet in the mud, in search of fresh water and food from the forests.

And the handmill? Well, it's still there on the wrecked deck of the ship on the seabed in Cardigan Bay, forever churning out salt, beer and mermaids, who are often mistaken by the environmental services for dolphins. And that's why a swim in the Salty Welsh Sea leaves you feeling as if you have been swimming with drunken mermaids.

Calan Gaeaf

It's Calan Gaeaf, the first of November. Mischief night, the first day of winter in the Welsh calendar, when the veil between this world and the Otherworld is at its thinnest and spirits from the past are able to pass through and haunt our dreams with scary stories of forgotten ancestors. A bit like a Welsh Day of the Dead, or Halloween when people wear bedsheets and pretend to be ghosts, or witches in green make-up with fake blood dripping from their mouths. Time isn't linear on this day. It travels in endless circles, swirling through space. You can travel around the coast of Wales in one day at any time in history.

Môrwen stands up, dripping mud on to the mossy forest floor. She remembers Nan's stories about the ice mountains melting and filling the valleys with floodwaters where her people live. She wonders if it will ever stop? And what happened to the mammoths? Did they drown? Môrwen can hunt like a wolf, swim like a mermaid and speak the language of birds, so she will escape the inundations – but what will happen to her people? She places a sprig of rowan in her pocket, faces the sea, sweeps back her hair, curses three times, and steps through the veil into her future.

Time passes in a moment.

Years, hundreds of them. Thousands. Millennia.

Môrwen hasn't enough fingers, toes, tails or limpet shells to count them all.

2

Siani Chickens and the Ceredigion Storycatcher

Myra the Storycatcher

Seven thousand years have passed in the blink of an octopus' eye, and Môrwen is on the same spot where she had stood after hauling herself out of the swamp, although she is now floating up to her nose in seawater. She swims through the jellyfish to a sandy beach, where a young woman with long black hair and a fringe that covers her sea-green eyes wraps her in a red and black Ceredigion quilt, and rubs her vigorously as if she were a toddler shivering after a cold bath. It's Calan Gaeaf.

'*Aw diolch,*' Môrwen's teeth chatter.

'*Croeso nôl, cariad,*' says the woman. 'I've been expecting you.'

'Where am I?'

'Llanina, halfway along the Welsh coast. Where better to start a fairy tale than in the middle? There are more stories here than anywhere else in Wales, and I should know because I've gathered them like cockles off the beach. Have some octopus salad. *Myra i fi.*'

Myra was born Elmira Rees in Cei Newydd, just before midnight on Calan Gaeaf, the first of November 1883, a child of the Otherworld, although the doctor arrived late and wrote her birth date as 2nd November, allowing her a foot in both worlds. Her father Thomas was Captain of the Rosina, a fisherman who lost all five brothers at sea. Myra wrote down the family fairy tales and stories told by the old sea captains of the town, and kept them in a biscuit tin beneath her bed so she could learn them while her father was away. They tell of the Otherworld, that dark enticing place filled with spirits, mermaids and *y bobl bach*, where you can vanish as easily as a small language beneath water and caravan parks.

'Tell me a story?' Môrwen asks Myra as she nibbles at the octopus salad, which is a tin of cheap luncheon meat cut into tentacle shapes.

Nodon's Well

A mean giant named Nodon lived at Llanina, next to a freshwater well covered with a slate slab that his daughter Merid slid aside whenever her father needed a bath, which was often. One day a gentleman rode up and demanded a drink. Merid explained it was her father's well, but the gentleman said water could not be owned, and he moved the slate slab and took a drink. The water overflowed and flooded the land, and as Merid was washed out to sea, she grew a beak and wings and turned into a gull. In time she gave birth to chicks, the ancestors of the gulls in Ceredigion today. And Nodon's well is still there on the seabed, forever pouring water into Cardigan Bay.

Chwedl Llanina

A fishergirl called Madlen lived with her father Gronw in a rush-floored cottage on the cliff at Traethgwyn, where they caught herring, crabs and lobsters and ate the straw from their mattresses, they were so poor.

One day, Madlen was locking up the chickens when she saw a bolt of lightning strike the mast of a ship flying a Saxon flag. Gronw had no love for the Saxons, but these were fellow fishermen, so father and daughter rowed out to the sinking ship, pulled seven men from the sea, and returned to save five more, although many others drowned.

Gronw and Madlen took the twelve survivors to their cottage and gave them warm mead while their clothes dried by the fire. One was a tall elegant man who spoke Anglo-Saxon, so Madlen fetched a linguistic monk from Henfynyw, who introduced the man as Ina, a Saxon King. In gratitude for saving his life, the King built a church on the spot where he was rescued, which he named after himself, Llan Ina. Though perhaps it should be Llan Madlen?

As the sea level rose the church was flooded, and a new one built on the clifftop, where it still stands today.

Fishergirl

As Môrwen and Myra tiptoe through the stranded jellyfish on the beach at Cei Bach, a fishergirl walks past on her way to sell a couple of lobsters to a pub in Cei Newydd where visitors demand freshly caught seafood. She once caught a beautiful electric-blue lobster, but the family who ate it never noticed the colour because it turned red when cooked. Fishergirl watched them eating, but they never noticed her either.

Siani Chickens

On the beach at Cei Bach in the early 1900s stands a ramshackle mud-walled cottage, with windows stuffed with rags and smoke steaming through the thinly thatched roof. In the doorway sits an old woman smoking a clay pipe, with a red shawl over her shoulders, a red and yellow spotted handkerchief tied round her face, and clogs on her feet. She looks for all the world like Mrs Tiggywinkle. This is Siani Chickens, who earns a penny or two telling fortunes and selling postcards of herself posing outside her cottage with her hens.

Siani waves her arms around like the hands of an old clock and shoos the girls inside her cottage. She sits Môrwen by the fire to dry her clothes in a fug of steam and makes a pot of Tregaron tea so thick she can stand her spoon in it. The cottage is full of chickens. Siani calls them *fy mhlant*, 'my children', and they all have names, Bidi, Ledi, Kit, Ruth, Beti, Marged, Charlotte, Cynddylan, and Jonathan the cockerel, who struts around as if he owns the place. On the table is a bowl of eggs dyed in Tregaron tea, which Siani sells from door to door. Everyone buys them because they love Siani, although no one can eat them because the eggs taste of saltwater.

When the shopkeeper in Llanarth was nasty to the poor children, Siani marched in and told him he was a miserable grumpy old man who should be kind to those less fortunate than himself.

Siani sings her favourite hymn, *Ar fôr tymhestlog teithio'r ryf*, while Myra warbles a verse of an old Catholic hymn, *Min Mair*, learned from Grandfather Daniel Williams of Glyngolau. Siani says she was born illegitimate, fled the workhouse, was poorly treated in love and ran away with the Welsh Romany. Myra laughs and tells Môrwen that Siani is really Jane Leonard, born in 1834 at Bannau Duon in Llanarth. She looked after her mam in Aberaeron until 1883, when she moved to the no-man's-land between the low tide and the cliff at Cei Bach, where she pays rent to no man. There are no boundaries in Siani's world, only the ebb and flow of the tide.

Môrwen asks, 'Aren't you lonely, Mrs Chickens?'

Siani laughs and points her pipe at the sea. 'No, no, no, there were five cottages here until the storms blew four down. Mrs Longcroft of Plas Llanina brings me food parcels, the *Bardd Gwlad* write poems about me, children visit to hear my stories, and Mr Môrgan visits me twice a day.'

Sometimes, Mr Môrgan bursts through her front door without invitation, so she climbs the ladder into the crog-loft and waits in her four-poster bed, sawn-off to fit under the rafters, with her chickens until he leaves. Mr Môrgan. *Y Môr*. The sea.

One day in 1917, Siani's neighbour on the clifftop noticed the smoke wasn't pouring through the roof. He found her lying peacefully in bed with her chickens and a biscuit tin containing £120 in threepenny bits with a note bequeathing her fortune to the poor children in Aberaeron workhouse. She is buried with her mother in Henfynyw Churchyard, and Mr Môrgan still visits Cei Bach every day in search of his beloved Siani Chickens.

Bwbach

Myra gives Môrwen a sketchbook and a few pencils so she can draw people on her travels. They will meet again in Swansea.

A hoar frost covers the ground. Skeletal spider's webs spread across the gorse. At Gilfach yr Halen, Owain Lawgoch sleeps in a cave waiting to save Wales in its hour of need. Môrwen wonders why he hasn't already woken up?

Calan Gaeaf swirls in all its weirdness around the pastel-sweet terraces of Aberaeron. A *bwbach* leaps out from under a bridge, sticks out its green furry tongue and bites through the rope carrying Captain John Evans' gondola across the harbour. The town is full of ghosts, apparitions, demons, skeletons, werewolves, zombies and vampires, and a few children in fancy dress carrying a giant mackerel. The spirits of the ancestors walk in the present. Môrwen should know. She is one of them.

Mesolithic Llanrhystud

The traffic noise hurts Môrwen's ears more than shrieking monsters, so she hurries down to the sea and paddles past the medieval fish traps off Aberarth, which supplied the monks at Strata Florida Abbey. At Llanrhystud, a forest stretched out to sea where she fished for salmon with a flint-topped spear along a mighty river formed by the Rheidol, Ystwyth, Dyfi, Leri and Wyre.

As she hops across the seaweed-covered tree stumps, the water level sinks and rises in front of her eyes. It turns from vapour to ice and back again, dogs transform into otters, falcons into black chickens, red deer drink at the water's edge, beavers build dams in the marsh, a forest of alder, pine, oak, hazel and beech grows from below the sand and rots into black peat. Time is whirlpooling and Môrwen feels seasick. She curses three times, wriggles her webbed toes, and everything calms.

An old woman sits on a log weaving a basket, a girl cuts the fur from a deer and another sings while sewing hides together to make a coat. Children dig up celandine tubers to cook and eat, while they bake eggs and fish on hot stones warmed in the fire. Others strip a felled tree trunk to make birch bark pitch, and thread periwinkle shells to make necklaces, while a skilled woman flakes an axe head with the sharp point of a stag's antler. She taps the flint and hears a ringing sound. It's good stone.

There are no villages or farms, just a few temporary shelters built with branches, animal hides, clods of earth and mud. These people are nomadic. They follow the seasons along the coastal plains, making homes wherever they go, chattering, gossiping and laughing while they work, and Môrwen understands every word. These are her people. This is her time. The land she left 7,000 years ago when she was Marsh Girl. A few moments ago, just then.

These travellers would not fit into our time. A law would settle them in affordable housing or static caravans, unable to travel without an address or a passport. Nomads need little proof of who they are, where they come from or where they are going.

Môrwen blinks and the forest fades to mist. She is still on the shingle beach at Llanrhystud. People in woolly hats and aprons are digging holes, scraping soil, measuring squares and scribbling on spreadsheets, as if searching for clues to her people's existence. One man drops a flint into a plastic bag.

Môrwen stretches her swan neck and peers over his shoulder. He is holding a tray marked 'Mesolithic', which contains more plastic bags filled with flints. She recognises one. Her mother used it to cut her daughter's hair. It was the sharpest tool they had. She tried a fringe but the flint cut her forehead. Anyway, bangs were so unfashionable in the Mesolithic.

'Hey, that's mine,' Môrwen tells the man.

He turns round and looks through her eyes at a woman behind her. He shows the contents of the bag. 'We think this was for cutting animal skins. It's Mesolithic.'

'No, it's for trimming hair. And it's mine. Give it back,' and Môrwen grabs it but her hand passes through as if she were a flickering shadow.

Môrwen asks, 'Is that me in the bag? Am I Mesolithic?'

She is invisible. They see her only through the objects they have unearthed.

'I can tell you flood stories. I have a name. It's Môr...'

They can't hear her. Is this all she is? A flint in a plastic bag? A watery memory from a forgotten folk tale? An Alice through a looking glass?

She climbs the cliff, past the caravan park at Morfa Bychan to Llanychaearn, where in 1826 a farmer watched a mermaid with a black tail from the cliff, head and shoulders out of the water, washing and throwing her dark hair behind her. It made the front page of *Y Cymro*: 'At first, he considered her modesty and then watched her for half an hour ...'

The water churns the sea white even on the calmest of days. She draws the archaeologist in Myra's sketchbook. Stories fly round her head like endlessly migrating birds as she walks north along the coast.

3

The Submerged Land and the Wise Old Toad

ABERYSTWYTH–BORTH–ABERDYFI

Mesolithic Tanybwlch

The early-morning beachcombers at Tanybwlch are tidying the plastic bottles, suspicious black bags and yellow rubber ducks washed up with sea potatoes, hornwrack, gutweed, razor clams, crab legs, mermaids' purses and a dried-out dogfish wrapped in orange rope.

In 1858 the mysterious Spanish barque *Tecla Carmen* beached here with no crew on board, apart from the ship's cat and all its lifeboats. It became known as the Ceredigion *Mary Celeste*.

A statue of the Duke of Wellington was meant to stand on top of the 10m-high column on Pen Dinas, but the idea was scrapped when the money ran out. The hill is haunted by a suitably headless dog, after the giant Cornipyn pulled too hard on its leash as he searched for his father Maelor, who had been kidnapped by mean mountain giants.

Môrwen sits on the limpet midden near the isolation hospital, where she and her friends told stories beneath the stars. They cooked herring in a pit covered with hot stones heated in a fire, gossiped about who kissed who, told comic tales 'til bellies ached, dreamed of sailing to the moon and terrified themselves with spooky stories until Nan calmed them with the dull legends of the tribe. The smell of cooking shellfish reminds Môrwen of Nan's voice.

The Ystwyth and Rheidol rivers meet here at the entrance to the harbour, bringing stories from the mountains to tell to the sea.

The Three Sisters

Three sisters rise high on Pumlumon, where curlews call and kites circle. Severn showers with spring water, smooths the creases from her business suit and flows east across the border, loops around Shrewsbury, under the iron bridge and on to Bristol. Wye rises next, pins up her hair, puts on her swimming costume under her jeans and T-shirt, tumbles her way over the rocks and stones, remonstrates with a farmer who is pouring slurry into her waters, stops in Hay-on-Wye to look at the books, and meets her sister Severn at Chepstow, where they run into the arms of their lover, *Y Môr*. Rheidol is still in her dreams – she throws on yesterday's clothes, drops the kids at school in Penllwyn, hurries through the industrial estate and the retail park, waves to the local otter and kingfisher, and on to Aberystwyth harbour, where she meets her half-sister Ystwyth to share chips and ice cream with the gulls on the prom.

Luna Park, Aberystwyth

At South Beach, Môrwen watches the pod of Cardigan Bay dolphins chase the silver mackerel close in to shore. She walks to the pier where starlings murmur each winter evening, past a busker who looks like Elvis Presley, and kicks the bar at the end of the prom, because that's what everyone does. Two elephants splashed in the sea near the bathing machines in the early 1900s, when the Cliff Railway carried passengers up Constitution Hill to Luna Park – a reproduction of the moon, rather like Georges Méliès' rocket in *La Voyage dans le Lune*. It went bankrupt a few years later, leaving only a white lady who prevents people from falling off the cliff edge.

Jolly the Smuggler

The glacial shingle spit of Sarn Gynfelyn stretches out to sea from the lime kiln at Wallog, thought to be the entrance to the submerged land and enchanted green islands in Cardigan Bay. Jolly the Trefechan smuggler took refuge here after escaping a gun battle with excise men at Tanybwlch. A cruel man who stole the coins from his dead grandmother's eyes, Jolly once killed and robbed a half-drowned sailor who had escaped from a shipwreck, only to discover it was his lost brother. When soldiers cornered him, he called on the Devil for help, and as a storm flooded the land, he leapt into the sea from Sarn Gynfelyn and vanished.

The Wise Old Toad

As Môrwen walks in a straight line through Borth, a wave leaps from the sea like a white horse, over the beach, across the road and floods the railway station. There's hardly any wind, only the pull of the moon and the late equinox tides. She wades through the water on to the station platform, past George Romary's railway museum-of-wonders, and across the track on to Borth Bog, watched by a nosey donkey. The once-extinct Rosy Marsh Moth lives on Cors Fochno alongside the rarest creature in Wales, the Wise Old Toad, with skin as leathery and blotched as the back of an old man's hand.

Toad sits, breathing and blinking.

'Are you the wise old toad? asks Môrwen.

Blinking and breathing.

'The oldest creature in Wales?'

Breathing and blinking.

'How old is the land?' asks Môrwen.

Blinking and breathi–

'Mesolithic girl?'

'Yes, but I have a name! It's Môr–'

'Speak up, I'm a little hard of *herring*. I'm so old I remember the first floods, when the earth we're standing on was as high as the highest mountains. I have eaten all the dust that filled the valleys of Wales, from the tops of the mountains down to this bog. Yet I've only eaten one grain of dust a day, as it wouldn't be sensible or sustainable to eat more. You can imagine how long that has taken? When you were born, I was already a grumpy, gloomy, wrinkled, miserable, teethgrindingly, bumscratchingly, mindbogglingly *ancient* old amphibian. That's how old I am.'

As Môrwen walks back to the sea, she hears Toad croak, 'Dimwit. Dimwit. Dimwit.'

Mesolithic Dyfi

Môrwen paddles through the submerged forest at low tide, now little more than dark stumps of oak, hazel and pine. She climbed these trees to stalk the deer herd, but the old stag usually outwitted her. The Mesolithic archaeologist from Llanrhystud stands in the sea holding her stag's antlers in the air, having pulled them from the peat below the sand. Môrwen places her foot in a bear print, and looks round in case she is being watched, but there are no bears in Borth now, unless one has escaped from the Animalarium.

The village is protected from flooding by groynes on the beach, made of green and purple heartwood imported from the Amazon in the 1930s, when the environmental cost of felling rainforests was unspoken. One day the floods will force people to search for sanctuary on higher ground by the church or in Tre Taliesin. It has been this way for so long, and no one would be stupid enough to make it worse by pumping sewage waste into the sea or warm the air with poisonous gases, would they?

The forest stretches towards the dunes at Ynys Las, where Môrwen waves to a girl who is practising handstands and cartwheels outside the visitor centre where her mam works. An osprey rises from the Dyfi estuary with a salmon between its claws, having flown here from West Africa to breed on a pole by the railway line. A low-flying military plane rumbles overhead, disturbing the egrets on the saltmarsh.

As Môrwen walks into the water, scales appear on her thighs, her legs become a tail, and she swims along the route of Edda Bell's ferry *The Hero* towards Aberdyfi.

Taliesin

Ceridwen lived in the forest at Llyn Tegid, where she brewed potions, set bones and drew out fevers. She gifted her giant husband Tegid Foel with enchanted armour so he would never be cut in battle. She could cure, charm and curse, so people called her a witch – which made her curse even more.

She saw herself as a mother with two children. Her daughter Creirfyw inherited her otherworldly skills and strength, while her son Morfran was cursed with his father's dark soul and brains. Ceridwen decided to gift Morfran with powers of prophecy and inspiration. With those, he would be a *dyn hysbys*, a conjurer.

By the light of the waning moon, she gathered herbs, leaves, bark, fungi, slime moulds, breath and odours. She built a fire from peat, and boiled the ingredients in a pitch-black iron cauldron. The potion had to simmer for a year and a day, so she employed Morfa the old blind beggar to sit and stir it, and Gwion Bach to pump the bellows to keep the fire alight, for he was a boy without a single thought in his head.

A year passed, and Ceridwen sat Morfran by the cauldron and told Gwion to give the bellows one last pump. There was a flame and a crack and the fire spat. Alarmed, Gwion stood up in front of Morfran and as he protected his face, the last three drops of potion burned the back of his hand. He licked it, and swallowed the three drops. Gwion had a thought. He had never had one before. His mind swirled with inspiration and prophecy. He called Ceridwen a witch, and ran out of the door as fast as a hare. Ceridwen screamed and gave chase as a greyhound.

As her tongue touched the hare's back leg, Gwion dived into the river and swam as a salmon. Ceridwen shrieked and gave chase as an otter.

As her teeth bit into the salmon's flesh, Gwion leapt out of the river and flew as a bird. Ceridwen squawked and gave chase as a hawk.

As her claws tore his feathers, he swooped into a barn and hid as a grain of wheat. Ceridwen would never find him there.

Ceridwen watched through the eyes of a jet black hen with a red comb, spotted the grain that was Gwion, and swallowed him down.

Ceridwen's belly grew and she gave birth to a baby boy. The child had a noble brow and bright eyes, but she knew he was Gwion's and she wanted rid of him before her husband suspected anything. She wrapped him in swaddling, dropped him in a leather bag, placed it inside a coracle, covered it with skins and set it free on the water at the mouth of the Dyfi river. It drifted on the tide. Gulls pecked it, gannets dived around it, fulmars spat at it.

A young fisherman named Elffyn caught the coracle in his seine net on the south bank of the river. He hoped the bag contained gold, so he could be the poet he had always dreamed of, but he unwrapped the swaddling to find a baby. Elffyn sighed, but the baby sang a poem, prophesying that he would be worth more than gold. Elffyn showed it to his father, Gwyddno, and the baby sang a poem so inspiring that it melted the old man's heart. So they raised him there on the banks of the Dyfi, where he grew to be Taliesin, the finest poet in the Welsh language, and repaid their kindness many times over.

The Submerged Land of Cantre'r Gwaelod

Off Aberdyfi, there was once a rich and fertile land criss-crossed with dykes and drainage ditches, sea walls, wells and sluice gates to keep the floodwaters at bay. Gwyddno's daughter Mererid defended the land against the sea while her father protected it from invading armies.

After one battle, Gwyddno threw a feast in honour of his friend, Seithennin, a mighty warrior who had fought many a war and drained many a glass. That night they celebrated with sweet mead, stuffed swans and pig heads, while Taliesin praised them

in poetry. Mererid had no time for such narcissism. She slept the stormy night alongside her sluice gates, preferring to keep the community safe from floods with science and engineering.

Seithennin, drunk on wine and blood, begged Mererid to cement the relationship between himself and her father. He plied her with drink and flattery, and as a storm swirled inside her, another gathered out at sea as the floodwaters breached her unprotected sea defences. The sea took the land Gwyddno thought he owned, and only he and Taliesin and a few others survived. In time, the submerged land became known as Cantre'r Gwaelod.

On Aberdyfi jetty is a bell that rings on the high tide to remember the flooding of Cantre'r Gwaelod. Songs fill the air, a memory of the last of the shantymen, Stan Hugill. Stan was born in Hoylake in 1906 and became a nautical artist, serving on the *Garthpool*, the last British commercial tall ship, which was wrecked in 1929 off the Cape Verde Islands. He spoke Japanese and many eastern languages, and wrote *Shanties from the Seven Seas*. He ran an outward bound school in Aberdyfi, and died in Aberystwyth, aged 86.

As Môrwen swims through the seaweed and barnacle-encrusted ruins of Gwyddno's submerged home, she sees another mermaid, with screws for earrings and a necklace of nails. As their tails twist around each other, she knows Mererid escaped the floods, too.

4

Migration Tales and the Cambrian Line

ABERDYFI–TONFANAU–ABERMAW/BARMOUTH

A Mesolithic Mermaid

A child watches Môrwen from the sand dunes between Aberdyfi and Tywyn, her eyes magnified by her glasses. She pulls her mother's hand. 'Mam, look, Ariel's waving at me.'

Mother fumbles in her bag. 'No, sweet pea. It's a seal. I think.'

The child opens her book of *Illustrated Welsh Folk Tales* and points to a painting of mermaids frolicking rowdily in Cardigan Bay.

'Mermaids are only in picture books,' says Mam. 'They're not real.'

The child studies Môrwen's tail, and wonders whether she believes her mother. She has had suspicions ever since the tooth fairy failed to honour the £5 invoice she pinned to her pillow when she lost a molar. Mam takes out her phone and searches 'Can seals wave?' but there's no wi-fi signal.

The child thinks Môrwen is a superhero.

The Floating King

Maelgwyn, King of Gwynedd, was about as popular as a swarm of mosquitoes in a swamp on a hot day. The people preferred wasps to him. So he hatched a plan to make himself more popular. He invited all the other kings to Aberdyfi for a competition, but without poetry or dancing like the 'Steddfod.

Maelgwyn placed thrones at the edge of the sea facing the incoming tide, and the last king sitting would be the most popular. Well, the tide flowed in and covered the kings' feet, their knees, bottoms, bellies and chins, but they refused to move until the sea drowned them for their arrogance. All except Maelgwyn, whose throne mysteriously floated like a coracle, for he had fitted pig bladders inflated with air to the chair legs.

So Maelgwyn proclaimed himself the most popular king in Gwynedd, for he was the only one, though the people loved Queen Nest far more than her overbearingly pompous husband.

Welsh Romany

Môrwen hauls out at the mouth of the Afon Dysynni, turns her tail into legs and walks to Craig Aderyn, Bird Rock, where the Wood family are having a Welsh Roma family gathering. The Woods are nomads like Môrwen's people. They don't own land. They belong to it. They camped on the Dysynni marshes before the valley was drained by German prisoners of war stationed here after the First World War.

Matthew Wood, storyteller and fiddler, closes his mystical deep-set eyes, breathes deeply, shakes his long black curls down his shoulders, and hails the birds in the Kale language as the words tumble from his lips like quicksilver.

The Frozen Ship

Choiya!

The captain of a sailing ship was looking for a crew when he saw Jack strolling along the quay with her long hair hidden beneath her cap, so he leaned over the rail and shouted, 'Ahoy, Jack, will you sign on as cabin boy for adventures on the high seas? You'll probably get eaten by a sea monster, but I will pay you handsomely in mouldy potatoes and stale beer.'

Captain explained that the King had offered a huge reward if he discovered a faraway wealthy land and returned with a gift worth more than gold. So Jack agreed to be cabin girl, for she loved nothing better than sailing round the world in search of exotic food.

Jack and the captain sailed across the ocean, and the weeks turned into months, until they were caught in a snowstorm that froze the sails and the ship ran aground on an iceberg. Jack scrambled to the top of the ice mountain, tossed a rope to the captain and hauled him up. They saw hundreds of steps leading down to a green fairytale land, so they climbed down and the further they walked, the hotter it became, and the ice melted.

They were welcomed by a farmer with food and drink and given a wagon to sleep in. Jack repaid the farmer's kindness by helping with the haymaking while the captain asked for a valuable gift to take back to his king. But there weren't any. The people were poor. They were travellers who had settled down to farm the land, and now they had little food or money. Jack understood, for she was a traveller too.

'You can have our giant,' said the farmer, 'he eats our potatoes and drinks all our beer. You'll have to catch him, though.'

So Jack lit a fire, cooked up a big pot of potatoes and left a barrel of beer by a tree. The giant smelled the food, tasted a potato and it was so good he ate the lot, then he drank the barrel of beer in one gulp and fell asleep in the warmth of the fire. Jack hauled the giant on to the ship's deck, propped him up against the mast and set sail for home with a cargo of enough potatoes and beer to keep the giant from eating her on the journey home.

By the time they arrived in Wales, Jack and the Giant were firm friends. So when the captain said he was going to give the giant to the King and collect his reward, Jack said, 'No, I caught the Giant. I'll take him.'

Well, Jack collected the reward, but the greedy King wanted more, so he said, 'Go back and get me some gold and take that smelly giant with you. He's eating all my potatoes and drinking my beer.'

So Jack gave half the reward to the captain in exchange for his ship, and set sail with the giant. They dropped anchor at the iceberg, climbed the ice mountain and down the steps into the green land, which was even poorer than before. So Jack gave the rest of her reward to the people who built wagons and travelled with the seasons once again. Jack grew potatoes and hops on the farm and taught the giant how to make his own chips and beer, which were far more valuable than gold.

And that's as close to a happy ever after as you're going to get. *Xolova!*

A Migration Tale

Môrwen places a cowrie shell in Matthew's hand, and he says she will see his family again before Calan Gaeaf is over. She walks past Sarn y Bwch, the second of the Sarnau reefs, where she smells potatoes cooking. She has never smelled anything so exotic before, so she follows her nose to the Tywyn fish and chip shop, where she sees a girl dressed in a sari.

'I love your dress,' says Môrwen.

'It's my sari. It's too cold to wear it outside, but it's warm in here,' says the girl.

'Where d'you live?'

'Tonfanau, though I came here from over the sea.'

'I live under the sea.'

'Then we are both migrants. I speak Hindi and Swahili.'

'I speak Welsh and Fish.'

They share a bag of chips and walk north along the stony beach to the old anti-aircraft artillery base with its disused aircraft hangars and derelict brick barracks.

'This is home. I arrived with my mother, father and *bapuji* from Kampala. They came to Uganda as migrants from India to build railways when it was a British colony. I miss the sound of rain pitter-pattering on our corrugated roof. After independence in 1962, the government told us we were taking jobs from Ugandans. So in 1972, thirty thousand of us were rounded up, many at gunpoint, given £50 and a suitcase, and told to leave in ninety days. I arrived at Heathrow Airport carrying a small blue suitcase on a freezing October night, dressed in my chiffon sari. Mother spent our money on a warm duffle coat and a bobble hat to stop me shivering. I looked like Paddington bear.

'We travelled on a train for six hours and arrived here at Tonfanau Railway Station with hundreds more. On the platform, a sign welcomed us: "Beware of the firing range." As I stared out to sea, the western wind nearly blew me back to Uganda.

'People came from all over with biscuits and cakes, and I was given a t-shirt and a denim mini-skirt. I was freezing. I snuggled up to the small electric heater in my room, hugged my suitcase, nibbled a biscuit and cried. I wasn't the same person I was a week before.

'That evening I wore my purple sari like a jewel under my duffle coat, and went for a walk by the sea. That's when I discovered the chip shop was as warm as Kampala.

'After six months, my family moved to a tiny brick-built semi-detached council house in Ely, where father found a job in the NHS. The pitter-patter of rain on our corrugated iron roof was replaced with enslaved housewives, bread riots and some people calling us names because they thought we were stealing their jobs.'

'So I decided to be myself. I opened my suitcase and dressed in my special sari with the colours of the Gunas who are present in all things. And ever since, my life has been a fairy tale. I am the girl who found home in a faraway land.'

Tonfanau is crumbling now. The plaster on one wall of the barracks has peeled away to reveal the shape of a rhinoceros. Môrwen catches the Cambrian Line train at Tonfanau station. The last train she caught took her to the moon, but this one is going to Pwllheli. She sits with her knees up to her chin and gazes through the window at the shapes of the round-stone walls, horizontal hawthorns and caravan sites. She has no money to buy a ticket. Fortunately the inspector sees only a shadow as the train clings to the vertical chicken-wired cliffs at Y Friog.

West Wales Railway Tales

On New Year's Day 1883, a train heading north collided with a rockfall at the Friog Cliffs and the engine fell 80ft into the sea below. Sailors were employed to rebuild the line, as their mast-climbing skills were better suited to vertical heights than railway gangs. Fifty years later, another train crashed into the sea, killing the driver and fireman. At that time, the train drivers met at Mr Jones' home in Pennal every Sunday evening to sing songs and tell stories.

The train circles Fairbourne, where houses huddle together awaiting the inevitable floods after the council gave up trying to repair the sea defences. There used to be a ferry to Barmouth here, but the train now trundles across the Mawddach on the longest timber bridge in Wales. The line almost closed in 1980, when the bridge was found to be riddled with marine woodworm. There are boats in the estuary stacked with lobster, crab and prawn pots, while people gather scallops, mussels and cockles on the mudflats. The summit of Cadair Idris is covered in mist. Spend a night on the mountain, the saying goes, you'll come down mad, dead or a poet – though some locals add, 'or an Englishman'.

Tegi, the Barmouth Monster

In 1937, Mr Jones of Harlech saw Tegi wandering along the riverbank. He described it as *Anghenfil y Bermo*, the Barmouth Monster: like a crocodile, about 10ft long, with a big head. A group of schoolgirls saw it in 1975 and said it had a long neck and a square face, black patchy skin and a tail with a flipper at the end. Tegi sometimes swims north, where it's mistaken for the Loch Ness Monster.

Afon, River

The train passes the sailors' institute, the lifeboat station, the dodgems and the Dragon Theatre on its way into Barmouth. Catherine Hutton, the solo traveller and diarist, sat on the beach here in 1796 with seven young girls, about 15 or 16, who took off their dresses down to their petticoats and went swimming. Three young men tried to steal their clothes, so Catherine stared menacingly at them while the girls dropped their wet petticoats and quickly dressed. They were the daughters of poor weavers, and after a day spinning cloth they wanted to swim and dance and flirt with the young men. Hiding their clothes was all part of the game, so Catherine left them to party, long before the arcades and funfairs were built.

Catherine noted in her diary that Barmouth folk called their river *Afon*, 'because it is the general Welsh name for all rivers. You are very fortunate if you find a person who can tell you it is the Maw.'

Môrwen understands this. Water is 'where the coral grows', or 'where a crab nipped my tail', or 'the green seaweedy supermarket trolley graveyard'.

Holden's Reef

She is tempted to go swimming on Chris Holden's Reef, formed by methane gas leaking from the seabed, where anemones, corals and crustaceans make homes on the soft rock. Môrwen loves poking her finger into coral.

The *Gossiping Guide to Wales* in 1881 says of Barmouth: 'The town itself has little to interest the visitor.' Which is not true. The Arousel Cafe attracts customers who understand the C has dropped off the sign above the Carousel Cafe door.

5

A Train Journey to the Mabinogi

The Cambrian Line

The Cambrian Line follows the sea through coastal fields criss-crossed by medieval drystone walls, where every wobble tells the story of the *pobl bach,* who denied the stonewaller access across their land. Through Morfa Dyffryn, with its orchids and nudists, towards Shell Island – which is more of a caravan site, airport and haunted farmhouse than a home for grooved razors and netted dog whelks. Near Llandanwg Station is St Tanwg's Church in the Sand, with a graveyard beneath the beach. On the hill above Llanfair are prints from the knees, feet and breasts of the Virgin Mary after she knelt for a drink of spring water. The algorithmic train announcer struggles to pronounce Dyffryn Ardudwy. It's raining. Drizzle that soaks.

The Bronze Bell, the Diamond and the Maid of Harlech

Offshore, Sarn Badrig – the largest of the three Sarnau reefs – is covered in a forest of bootlace weed, sugar kelp, podweed, coral and shipwrecks. The *Bronze Bell* broke up here under the weight of its cargo of forty-three blocks of Carrara marble. The ship's bell, dated 1677, bore the inscription *Laudate dominus omnes gente*. 'All people praise the Lord'.

The *Diamond* was built in Manhattan for the Macy family who ran the famous department store. On the night of 2nd January 1825, the ship captained by Henry Macy ran aground on Sarn Badrig with a cargo of American apples and wealthy cotton barons. Only nine people survived, along with the apples, which were made into pies and the pips planted in gardens. In the early 2000s, plantsman Ian Sturrock discovered an old tree with bright red fruit in a garden at Dyffryn Ardudwy, so he grafted its wood on to a tree in his Bangor nursery and named the new variety Diamond.

Occasionally visible at low tides is the ghostly shape of *The Maid of Harlech*, a Lockheed P-38 Lightning that crashed on the reef in 1942 after taking off from Llanbedr on a gunnery test. In 1988, a 100-year-old leatherback turtle was caught in fishing lines and washed up in this shallow coral graveyard.

Branwen

Nothing is happening. No announcements. Maybe the driver is waiting to pass the key to the southbound train to allow access onto the single-track line north? Coleg Harlech slowly crumbles on the cliff along with its aspirations. Castell Harlech glowers from the top of Ffordd Pen Llech, once officially the steepest road in the world until Dunedin's Baldwin Street was thought to be steeper. Outside the castle is Ivor Roberts-Jones' sculpture of Bran, the broken King of the Brythons, who sits on his horse with his dead nephew Gwern behind him. This is Mab World.

A voice asks if the seat opposite is taken. Môrwen scowls as a girl sits down, plugs in her mobile phone and offers her some single-origin dark chocolate buttons. She has dark hair in a plaited ponytail, and is wearing a blood-red coat with a starling perched on her shoulder. She has stepped through the veil on Calan Gaeaf, too.

'I like your bird.' Môrwen stuffs a handful of buttons in her mouth.

'She's my messenger bird,' says the girl. 'I know WhatsApp is quicker, but I can't part with her. I'm Branwen ferch Llŷr, sister to Bran, once King of this disunited kingdom. You might have read my story in the Second Branch of Y Mabinogi.'

'I can't read. I like stories with pictures,' Môrwen draws Branwen's starling in her sketchbook.

'I'll paint stories in your mind, babe,' says Branwen, 'you can sketch them.'

Branwen begins, 'I lived here in Harlech with my brother Bran and half-brothers Nisien and Efnisien. War with Ireland was in the air, so Bran invited Matholwch the Irish King for peace talks, and ordered me to sleep with him to unite our countries. I told him no way, it was my decision who I slept with, but after a night of feasting, I found myself in Matholwch's bed. While we slept, my warmongering half-brother Efnisien slaughtered the Irish horses. In the morning, Matholwch saw the bloodshed and raised his sword, but Bran calmed him with the gift of a *Pair Dadeni*, a cast-iron cauldron that restored life to those who died in battle.

'Matholwch and I sailed for Ireland in thirteen ships with the cauldron on board. The Irish people treated me like a princess, and I had a son, Gwern, but my husband gave him to foster parents to raise as a warrior. Heartbroken, I told him I wanted my boy back, but his people laughed at him for marrying a stroppy Welsh girl, so he sent me to the kitchens and ordered the butcher to slap my face each day with bloodied hands!

'For three years, I refused to speak with men. My only conversation was with this sweet starling who sang from the kitchen windowsill. I poured out my heart in a letter to Bran, tied it to her wing and she flew to Wales, where she found my brother staring out to sea near Harlech, before the sea flooded the dunes.

'When Bran heard of my husband's cruelty, he called his warriors, and after a feast they dressed in their armour and marched towards Ireland. With his harpers at his shoulders, he waded through swamps and marshes, across two rivers, Lli and Archan, and towed his ships across the Irish sea. The King of Ireland's pig-keepers saw what they thought was a moving mountain covered in trees, so they alerted Matholwch.

'He fetched me from the kitchen and asked, "Branwen, what is this?"

'"How would I know? I'm the kitchen maid you threw from your bed. Maybe it's an army come to rescue me?"

'"What are the trees?"

'"Masts of ships."

'"What is the mountain?"

'"My brother the King. There's no ship big enough to hold him."

'"And that ridge and those two lakes?"

'"His nose and eyes."

'Matholwch realised an invading army was on the horizon.

'He retreated over the Shannon, and ordered the bridge to be burned behind him. So Bran built another bridge with his body and his warriors laid wooden planks over his back to cross over.

'Matholwch offered to compensate Bran by making my son, Gwern, King of Ireland. I advised my brother to accept, for I couldn't bear the thought of my two countries at war. They called a peace council, but Matholwch played a trick. He hung bellies, skin bags, from posts in the meeting hall and hid armed soldiers inside each one, 200 in all, but my half-brother Efnisien wanted blood, so he silently slew them all.

Mesolithic girl takes the first selfie of a Mesolithic girl.

'My son Gwern was crowned King of Ireland, and the warriors praised him, all except Efnisien. Gwern saw into his uncle's dark soul, so Efnisien grabbed his feet and hurled him head first into the fire. I leapt towards the flames to save him, but the Irish warriors drew their swords, and Bran held me between his shield and shoulder. As my son burned, there was a terrible slaughter.'

'The bodies of the Irish dead were stripped to the waist and thrown into the *Pair Dadeni*. Soon the cauldron overflowed with reborn warriors until Efnisien shattered it into pieces and with his last redemptive breath ended the war. We were victorious - if victory it was, for only seven survived: Pryderi, Taliesin, Manawydan, Gilfiau, Ynawg, Gruddiau, Heilyn. Bran was pierced in the foot by a poisoned spear, so he ordered his head to be chopped off before the poison reached his brain. His talking head was taken to Gwales, Grassholm Island in Pembrokeshire, and I'll tell you the rest when your journey takes you there.'

The train circles round Traeth Bach, where a flock of Canada geese graze on the salt marsh. The gorse is in flower, which means it's kissing time. There are three species of gorse and their flowering overlaps, so it's always kissing time up north. Môrwen presses her lips to Branwen's forehead. There are no words. These horrors are the same the world over, and have been throughout time.

Branwen places her phone in Môrwen's hand, 'Here, honey, I've bookmarked a map. Visit my grave on the banks of the Alaw when you get to Ynys Môn.'

They take the first-and-only selfie of a Mesolithic girl and a Mabinogi girl, before Branwen leaves the train at Penrhyndeudraeth.

The Moth Who Caught the Train

In 1965, a Mr Revell, an entomologist, was on holiday by the sea near Penrhyndeudraeth station when he caught a Rosy Marsh Moth, *Coenophila subrosea*, in his moth trap. This was odd, because Rosy hadn't been seen in over 100 years and was believed to be extinct in the UK. So a survey was carried out to find Rosy's food plant, Bog Myrtle. It transpired the nearest plants were on Cors Fochno, 100km from Penrhyndeudraeth. Two years later, 192 caterpillars were found there. There was no other explanation: Rosy caught the train from Borth and got off at Penrhyndeudraeth station, and, like Môrwen, never bought a ticket.

A UFO at Portmeirion

At Minffordd station a sign welcomes people to Portmeirion, the Italianate fairytale village designed by Clough Williams-Ellis in the 1920s. Shortly before it was built, Jenny, the daughter of the Reverend Jones of Penrhyn Isaf, was walking home from Penrhyndeudraeth followed by her father's servant Dafydd Fawr, who was carrying a large flitch of bacon. Jenny brushed the brambles and nettles aside with her skirt so Dafydd could walk unscratched, but as they approached Aber la she turned round to find he'd vanished. She assumed he had sloped off to the pub in Minffordd, so she continued home alone.

Dafydd was behind her when he saw a meteor fall to earth in the woods. A small man and woman stood in the middle of a hoop, holding hands with their feet resting on the ring. They jumped out and danced in a circle on the ground to the sound of the sweetest music. Then they jumped back into the hoop and flew away, so Dafydd followed Jenny home. She asked him where he had been, because she'd been home for three hours. He said he'd seen a UFO.

She didn't believe him. Obviously he'd been dancing in a fairy ring, like most young men in Wales.

Sioni Onions

Môrwen leaves the train in Porthmadog and wanders through town past Browser's Bookshop to the Cob, which carries the road and the Ffestiniog Railway across the estuary. She steps into Cob Records, the legendary vinyl shop opened in 1967 to sell ex-jukebox singles, and now stocks 15,000 quality rare albums. Môrwen is torn between Datblygu's *Peel Sessions* on Ankst, Georgia Ruth's *Mai* and Cowbois Rhos Botwnnog's *Red Vinyl*. She has no money, so the manager gives her a sticker that reads, 'Cerys Havana Didn't Kill Nansi Richards'. She sticks it on her arm and pretends it's a tattoo.

Outside the shop, Sioni Onions is waiting for her with his bicycle. He is one of many Breton men who cycle around Arfon and Dwyfor selling strings of onions. They work for a company in Porthmadog run by Claude Deridan and his father. Claude was born in Roscoff in 1904, when over a thousand Sionis worked in Wales. He learned Welsh, marched with the veterans on Armistice Day, and was well-loved in the town.

Môrwen climbs on to Sioni's spare bike and she wobbles through Criccieth to Pwllheli Harbour, where an eel has wrapped itself round a heron's beak to avoid being eaten. They follow the Tramway to the Polish-speaking village of Penrhos, a former airfield settled by war veterans and refugees from Poland after the Second World War.

At Felin Uchaf, carpenter Dafydd Storïwr has built a wooden village with a roundhouse theatre, where he tells a local tall tale:

Donkey's Ears

March, the wealthy King of Pen Llŷn, had a secret. He had long furry donkey's ears. He grew his hair into an elaborate beehive to hide his ears so his subjects wouldn't laugh at him, but his hairdresser knew the King's secret. Like all secrets, it gnawed away at her, until she simply had to share it or she would burst. So she went to the river and told it to the reeds. A passing musician cut one reed to make a set of pipes to play at a party at March's castle, but when he blew on the pipes, they sang, 'March has donkeys ears', over and over again, and soon the whole of Pen Llŷn knew the vain King's secret.

There used to be a cafe in Abersoch close to Castell March with murals depicting scenes from the story. When the cafe was demolished a few years ago, the paintings were dumped. A shame, because there are still plenty of Marches in Abersoch, though their donkey's ears are hidden by fashionable yachts, second homes and crazily priced beach huts.

6

Island Tales of the Kings of Bardsey

ABERDARON–YNYS ENLLI/BARDSEY ISLAND

The House with the Front Door at the Back

An old farmer with a stubbly chin lived in a damp house at the end of a row of cottages between the river and the sea in Aberdaron. Every morning he padded down the creaky stairs in his socks, stepped outside the front door, dropped his trousers, and did a poo by his doorstep. Well, this was long before houses had a *tŷ bach*, never mind a bathroom. And the muck heap was great for mulching potatoes.

One evening he sat there, trousers round his ankles, when he realised he was being watched by two little people covered from head to toe in poo.

The woman spoke, 'We lived here long before you were born, and yet every evening, we sit by our fireside and poo pours down our chimney.'

The man added, 'Our baby is covered in it. If you don't believe me, step on my foot three times.'

So the old farmer stepped on the little man's foot, and there, next to his front doorstep was a tiny house, and as far as his eyes could see was a village with cobblestone streets and people dressed in strange fashions, walking around as if he wasn't there. Herds of white cattle and enchanted pigs fed on scraps in every alleyway. The farmer knew he had passed through the veil into the Otherworld of the *pobl bach*.

He blinked, and the town vanished.

The farmer was mortified. He never meant to hurt the little people, so he unscrewed his front door, piled stones in the doorway and covered it with lime mortar, so you would never have known there had been a door there. He planted his muck heap with sweet-smelling evening primroses and night-scented stock, then he knocked a hole in his back wall, screwed the door in place, and crossed his fingers that no little people were living by his back doorstep.

There probably were, though.

Dic Aberdaron

Richard Robert Jones was born in 1780 in Aberdaron, where he refused to go to school or become a fisherman like his father, because he wanted to be a poet and musician. So he walked barefoot along the old Welsh Tramping Roads in a blue poacher's coat with pockets full of books, a hareskin hat on his head, ram's horn round his neck, harp on his back, and a cat on his shoulder. He sang the songs of Homer on the streets of Liverpool and sold his books of poetry to buy food. He claimed to speak thirty languages, though only half fluently, and he spent two years compiling a Greek–Hebrew–Welsh dictionary but never found enough subscribers to pay for publication. He owned a Book of Spells and could conjure pig-demons called 'Cornelius' Cats', which he once invoked to help a farmer cut hay. Artists painted his portrait and fellow bards wrote odes about him, though one said that Dic was three quarters genius and one quarter idiot. His dictionary was buried with him, while his book of poetry is archived in the National Library of Wales, a fitting resting place for a Welsh folk hero, still remembered while more 'important' men are long forgotten.

The Preacher Poet

Môrwen is eating a raspberry ice cream outside Y Gegin Fach next to the seventeenth-century tea room Y Gegin Fawr, watching a little boy catch eels in the River Daron with a net and bucket. He's been banned from mam's kitchen since the eels learned how to escape.

She leaves the bike for Sioni Onions, and sits on the beach with her back to St Hywyn's church wall. She scoops up a handful of sand to find a Roman coin in her palm. She rubs it between her fingers like a magic lamp. It must have come from the graveyard. She tries again and finds a small ceramic lobster from Mary's gift shop.

She follows the lane to Porth Meudwy, where the Bardsey Island boat leaves from. She meets a tall man with a pair of binoculars round his neck who talks animatedly about the warblers he saw in the spring. He is poet R.S. Thomas, vicar of St Hywyn's, who moved to Aberdaron in 1967 to enjoy solitude and birdwatching. He advises Môrwen to take only bread and cheese to Bardsey and live simply, just as his predecessor St Hywyn preached from a wooden cell.

The Bardsey Boat

In the late 1800s, the antiquarian Thomas Pennant, on a story-gathering trip round Wales, observed the Bardsey boys stop rowing halfway to the island, 'tinctured by the piety of the place'. The old wooden beach-boat towed a smaller boat or two loaded with supplies of vegetables, tin cans, and coal bags, which bobbed up and down through the tide race around the Devil's Ridge and Ship's Ledge. In Aberdaron churchyard is a gravestone that commemorates Thomas Williams and his daughter Sydney, two of six people who drowned when the Bardsey boat was wrecked in a storm on 30th November 1822. Evan Pritchard, bardic name Ieuan Llwyd, wrote 'The Bardsey Boat Lament':

I heard a shout over the deep salt water
It was a cry full of sadness from Bardsey
A shout above the roar of a cruel wind
And the great tumult of the sea
The shout of widows and orphans
A throng of the weak: the lives
Of husbands and fathers ended in the waves
On the eddies, rocks and cross-currents

Pobl Enlli

At Porth Meudwy the boat is filled with supplies for the island. There are no day visitors, so Colin the boatman teaches Môrwen how to drive. A hundred resident seals sing the boat on to the slipway at Cafn, where Moss the farm dog greets her. He will stay by her side while she is on the island.

The remains of one of the old wooden rowing boats lies upside-down, surrounded by old lobster creels and crab pots. There are still plenty of lobsters in the nooks and crannies around the island, but edible crabs are scarce thanks to industrial trawlers scraping them up from the seabed three miles offshore.

Moss races Môrwen to Solfach, where her people left stories of their lives. There are horizontal lines filled with flints in the crumbling cliff, a fossil record of meals, fires and floods. She once walked here from Llanrhystud with her father through swamp forest and salt marsh. They rowed the last few kilometres across the tide race in a hollowed-out log canoe. She built a fairy tale castle out of limpets and spider crab legs. One day, archaeologists will unearth Marsh Girl's castle and call it a midden.

She follows the island's single track past the farm where the island poet feeds the lambs, chickens and turkeys throughout spring, summer and autumn. She will write when she returns to the mainland before winter arrives. There is too much to do on Enlli to find time for literature.

They pass the Hen Ysgol, where generations of children were taught about island life; past the Bird Observatory, where R.S. Thomas ate his bread and cheese; and Plas Bach, where a Bardsey apple tree likely grew from the pips in a discarded apple core. Most of the houses have no electricity or bathrooms, just solar torches and buckets, so Môrwen pays a visit to Andy's luxury compost loo.

As they walk towards the north end, the twenty-first century is left behind. Ynys Enlli is an island of 20,000 Saints and 10,000 Manx shearwaters, a mythical mix of sacred and secular, where pirates built houses with stones from the ruins of St Mary's Abbey, and human bones have been unearthed when ploughing. Fishermen, farmers, artists, poets and naturalists cling to the land by their fingernails while Myrddin sleeps in a glass tomb beneath the abbey.

In the 1800s the Islanders were fiercely independent. They lived on salted meat and fish, oatmeal porridge, bread made from island barley and beer brewed from home-grown hops. They wove cloth and dipped rush candles in fat for light, kept cattle, pigs, chickens and horses, farmed the sea for crabs and lobsters, and gathered driftwood to build boats. They quarrelled, shared stories and once feasted on oranges and bananas gathered from a shipwreck.

Tomos o Enlli tells of a farmer walking across the mountain one evening when he heard singing like silver bells and felt someone alongside him. When he reached *Ogof y Tylwyth Teg* he saw a shining girl with golden hair. She disappeared into the cave and the farmer couldn't remember how he got home that night.

The smugglers were co-ordinated by Lord Newborough of Pen Llŷn, but the wreckers worked only on the mainland. A fisherman and his son once saw a line of hooded figures carrying lanterns near Aberdaron, so they rowed back to the island knowing a storm was gathering and a ship would be wrecked.

Early in the Napoleonic wars, Siôn Robert Griffith, the only English speaker on the island, was paid by the press gangs to entice the rest of the men on board their ship. When the women noticed their men were missing, they called Lord Newborough, the ship was intercepted and the men brought home, apart from Siôn, who used his blood money to reach America.

The Kings of Bardsey

In 1826, the islanders elected 27-year-old John Williams, who farmed 12 acres at Cristin Uchaf, as their king. At his inauguration he stood on a chair at Cafn in the presence of the lighthouse keeper, while Lord Newborough presented him with a tin crown for dignity, a silver snuff box for wealth, a wooden soldier for an army, and ribbons in his hat for authority. Fifteen years later, King John drowned while crossing Bardsey Sound alone, and the following day his son John was born.

King John II grew to be a gloomy drunkard. After thirty-six years on the throne, he was advised to abdicate. When he refused, the islanders staged a coup. When the King was drunk, they rowed him to Aberdaron and dumped him on the beach for the mainlanders to deal with.

His successor Love Pritchard, farmer and fisherman from Tŷ Pellaf, inherited the crown with the words, 'I am the oldest, I am going to be King now.' In 1914 Love offered to fight for the King of England, but was politely rejected on the grounds that he was 71. Love took offence, and declared Bardsey neutral. In 1924, with life on the island increasingly impossible through discontent and hunger, he led an exodus to the mainland. On a visit to Liverpool, a waitress asked him, 'What would you like, love?' and with tongue firmly in cheek, he said, 'See, they know me here.' When he turned 83, he still hoped to find a queen, but when he died the monarchy passed with him.

Island Folk

Time passes slowly. There aren't enough hours in the day. A chough flies over Hendy. Manx shearwaters turn the sky dark. Gannets dive into shoals of fish. A pod of Risso's dolphins breach the water, each with its own uniquely scarred and shaped dorsal fin. They were first identified by the partner of the warden at the bird observatory in the 1990s, but few listened to her then. A minke whale is spotted off the North End, while a bullock from Lundy Island has washed up on the beach at Henllwyn and is ritually burned. The fire takes four days to dampen down.

There is a need to understand the sea's moods, sense the approaching weather, watch the stars, know when a sick lamb must be taken to the vet in Pwllheli, and understand the humour of remoteness. In the 1950s Brenda Chamberlain, author of *Tide-race*, sketched horses on the walls of her home at Carreg Fawr, until a pious visitor rubbed out their genitals. Amelia Shaw-Hastings documented the island and people for twenty-years with drawings of Wil the boatman; Kim Atkinson, the illustrator of *Tomos o Enlli*, and the lobster fisherman's dog, Fly. Gwilym Pritchard and Claudia Williams stayed at Nant and painted each day. The island is a fantasy seen through the eyes of artists and children.

Môrwen draws a lobster in her sketchbook while Moss runs off to fetch the farmer, who gives her a lift on his quad bike to Cafn to catch the boat. A thousand and one fairytales and a couple of Bardsey apples leave with her to guide her back to the Otherworld.

As she looks across Bardsey Sound, she imagines Bardsey and Aberdaron were once in love, and when they argued the island shook the earth beneath the sea and moved herself away from him. The tide race is a memory of their enduring passion. On 19 July 1984, an earthquake measuring 5.4 centred off Pen Llŷn was felt in Liverpool and Dublin.

7

Rockpools and a Giants' Town on Pen Llŷn

Llawer o Wisgu (Whisky Galore)

The barque *Stuart* was on its way from Liverpool to Wellington in New Zealand with a cargo of whisky when it ran aground at Traeth Penllech in April 1901. Before the customs men arrived, the whisky was hidden in caves, barns and down rabbit holes by the happy people of Llangwnadl. Only one bottle was ever recovered, and that was 110 years later. Wellington may have lost its Welsh whisky, but they have the consolation of being able to buy Penderyn Single Malt from the refurbished city urinal, which is now the Welsh Dragon Bar.

Tylwyth Teg

On the narrow peninsula at Porthdinllaen is an entrance to the Otherworld, home of the *tylwyth teg*, who are only seen when the weather is misty, which is most of the time. A golf course was built over it in the early 1930s, since when the little people have supplemented their income by selling lost balls back to the clubhouse. Their mischief may be the reason why developers rejected Porthdinllaen as the ferry port for Ireland and chose Holyhead instead.

A girl who lived on the beach at Nefyn played for hours on the cliff west of the town with some children, who she thought were much nicer than she was. One day, her mother followed the girl to the clifftop, where she introduced her friends. Mother saw nothing, and told her daughter never to visit the *tylwyth teg* again.

Lowri Hughes' nan from Nefyn was milking a cow at Garn Boduan when a dog came sniffing around. Nan kicked it away and it ran off, whining. It returned with a lame fiddler who asked for milk, but Nan refused. The fiddler began to play, and Nan found herself caught in an everlasting dance with the *tylwyth teg*, tormented forever by her own cruelty.

Griffith Griffith from Edern was walking to Caernarfon at two in the morning to pay his rent when he saw a crowd of little people coming towards him, speaking a language that wasn't Welsh or English or anything he recognised. He waited respectfully by the ditch while they passed, for he knew they were *tylwyth teg*.

Eddie Kenrick of Edern began gathering local stories from books and oral history after he was severely wounded in France in 1917. He became a long-distance runner and freelance journalist, and printed his stories on rolled up pink photocopy paper to give to anyone interested in folklore.

Rockpools

Môrwen leaves a few yellow golf balls as a donation to the *tylwyth teg* for crossing their land. She sits by a rockpool on Porthdinllaen beach and watches a red octopus crawl out and scamper into the sea. These rockpools are underwater museums for children, archives of seaweed forests, limpet houses and mermaids' purses. As she draws the octopus in her sketchbook, Jane Jones, landlady of the Tŷ Coch and first woman harbourmaster in the UK, brings her a Nefyn beer on the house.

Cottages huddle beneath the cliff at Porthdinllaen, once a thriving fishing community that traded in salt and herring with

Ireland and Liverpool. Carpenters walked from Llaniestyn to work on the beach from 6 a.m. to 8 p.m. each day, where they built ships like *Fanny Beck*, *Miss Wandless* and the 20-ton *Mermaid*. Evan Hughes of Pen yr Orsedd carved figureheads, while David Rice the moneylender ran the lime kilns. The Tŷ Coch was the centre of the world, along with the lifeboat.

On 20 September 1975, Coxswain Griff Jones of Morfa Nefyn was on leave when the lifeboat *Kathleen Mary* was called to search for survivors from a yacht lost in gale force winds. As he drove through the golf course he saw a man clinging to a rock near the boathouse, illuminated in the car headlights. Griff and his 14-year-old son Eric rescued the man in the boarding boat and received a bronze medal in recognition of the RNLI's kindness to strangers.

The Nefyn Mermaid

A young fisherman sat on the beach at Nefyn, mending his nets and watching the gannets diving into a shoal of herring, when he saw a mermaid on the tideline. Unlike the sirens in picture books, she had limpets on her shoulders, barnacles dangling from her ears, and crabs crawling over her tail. He enticed her with words to his cottage on the beach, for these Pen Llŷn fishermen are silver-tongued poets, although as time passed he failed to see she was drowning in his obsession. She pleaded for saltwater and sang a lament for the sea, so he placed the mermaid in a rockpool. As she rode away on the white horses, she promised to warn the fisherman of shoals of herring:

Penwaig Nefyn, penwaig Nefyn,
Boliau fel tafarnwyr, cefnau fel ffarmwrs.

Nefyn herrings, Nefyn herrings,
Bellies like inn keepers, backs like farmers.

The fishermen were known as *Penwaig Nefyn*. In 1907, a Pwllheli fishmonger sued another who dared to sell fish from Norway as *Penwaig Nefyn*.

Ifan Morgan, another poor fisherman, found a mermaid eating his fish in the cave where he stored his nets. She told him her name was Nefyn, and she was in love with his singing. She showed Ifan a stash of gold and gave him a fisherman's cap as a love token. Well, Ifan couldn't resist gold, so he invited the lovelorn mermaid to live in his cottage, where they had five children. One day the family were out in Ifan's boat when Nefyn leaned over the side and asked the sea for calm weather. This was the moment her children knew their mother was a storm whisperer, a mermaid.

Another *Penwaig Nefyn* became infatuated with an elegant London woman half his age who had taken refuge from the material world in a cottage near Pistyll. When she returned to her theatrical life in Camden Town, he followed to find she was married to a depressed art historian from Tufnell Park. Too ashamed to return to Nefyn, he worked as a refuse collector in Soho for the rest of his sad days.

The Corpse Bride

Meinir's nan heard singing outside the family farm, Tŷ Hen, in Nant Gwrtheyrn, followed by a knock on the door. She opened it to find young Rhys come to take his cousin Meinir to St Bueno's church in Clynnog, where they were to be wed. Nan invited him in, but all he saw was an old woman in a stick chair wearing a long black dress with a high lace collar, a pair of spectacles balanced on the end of her nose and lines on her flour-white face as if drawn with charcoal from the fireplace.

Nan told him to look for Meinir under the beds, in the *tŷ bach* and the pigsty, but there was no sign of her, so he searched the schoolroom, the chapel and the bracken on the cliffs. When he was out of sight, Nan scrubbed the flour and charcoal from

Meinir's face, removed the black dress to reveal a white wedding gown, pushed her out the back door, and told her to run away to somewhere exotic like Paris, Rome, or Aberystwyth.

Rhys knew this hide-and-seek was part of the horse wedding tradition in Y Nant, but as the days, weeks and months passed, Meinir was nowhere to be found. Some said the *tylwyth teg* had taken her, but Nan knew it was the Curse.

You see, three monks from Clynnog once visited Y Nant to preach God's word, and as the locals chased them away, they cast a spell – if any cousins dared to marry, the village would slowly die before being reborn. Rhys didn't believe in these fairy tales, and knew Meinir was waiting for him to find her, so he continued to search with his dog by his side until, during a fierce storm, a bolt of lightning struck a hollow oak at the foot of Yr Eifl, and as it split open, a figure in a white wedding gown floated away.

The lovers are still seen, walking arm in arm along the beach towards Carreg y Llam, he with long white hair and she in a frayed wedding dress, forever the corpse bride and groom of Nant Gwrtheyrn.

The Village Co-op

And old Nan's story of the Curse? Well, the village did die when the quarry closed in 1964, leaving people without work and the family houses in Nant Gwrtheyrn empty. The coastal villages were becoming retirement homes and holiday camps, and young people were leaving to find jobs in cities or escape to art colleges.

Until in 1970, Dr Carl Clowes and his wife Dorothi moved to Bryn Meddyg in Llanaelhaearn, and discovered the medical issues in the village were linked to the social, economic and political problems. After much discussion in Tafarn y Fic and Tafarn Yr Eifl, Carl helped found Antur Aelhaearn, the first community co-operative in the UK. Locals bought shares for £1, and the funds allowed youngsters to learn new skills in the pottery and knitwear factory. Nant Gwrtheyrn was later rebuilt as the National Welsh Language Centre and the land was reborn, exactly as Nan's Curse foretold.

Legends and real life are intertwined here. Two of Carl and Dorothi's children, Dafydd Ieuan and Cian Ciarán, founded the legendary band Super Furry Animals with Gruff Rhys.

Swan Girl

Close to Llanaelhaearn is Llyn Glasfryn, where Grasi worked as the well-keeper, a job so dreary she dreamed herself into fairytales. One day, she was flying to the moon on the back of a swan, when she forgot to close the well. The waters overflowed into Llyn Glasfryn, and as Grasi was swallowed up by the lake, her neck stretched, her arms sprouted feathers, her nose turned orange, she hissed and spat and transformed into a swan. For three hundred years Grasi swam round the lake, and every morning she was heard whooping and weeping, which frightened the servants at Glasfryn House. After she died, she haunted the lake dressed in a spectral white gown, which terrorised the people even more.

A Changeling Child

Môrwen climbs through the heather and gorse on the shoulder of Yr Eifl, where a giant man with a chough on his shoulder is scything the bracken.

'What's her name?' asks Môrwen as she strokes the chough.

'Brân Goesgoch. She has red legs, see?' says Dan.

'Where did you find her?'

'She was hatched in the Nant quarry where I worked. When I lost my job, she followed me. She is keeping me company while I clear the bracken for Jonny Brynffynon so his sheep can graze the common land.'

Dan gives Môrwen some grubs to feed to the chough, and they sit on a rock while he tells a tale.

'Elis Bach of Tŷ Canol, Nant Gwrtheyrn was a changeling child. All his brothers and sisters were the size of humans, but Elis had legs so short his body almost touched the ground. Yet he could run like a hare, and with his dog, Meg, they rounded up the mountain sheep for the Nant livestock markets. One day, Elis saw two men offering to pay over the odds for lambs. So his mam invited them in for cawl, while Elis hid in a cupboard and listened to the men discussing sheep stealing. Later, Elis saw them herding stolen lambs up the corkscrew road, so he took a short-cut through the trees, waited at one of the bends in the road, leapt out in front of them, did a weird dance, frightened the lives out of them and chased them to Pistyll, while Meg led the sheep back to the Nant. You see, *tylwyth teg* are no different to people on Pen Llŷn.'

Giants' Town

Môrwen waves to Dan and Brân and clambers up Tre'r Ceiri, the Giants' Town, to the remains of 150 Iron Age stone roundhouses on the summit. From here she can see the whole world spread out like a map. Well, maybe not all of it, but Pembrokeshire in the south, Snowdonia to the east, the English Lake District up north, the Mull of Kintyre in faraway Scotland, and the Wicklow Hills in Ireland to the west. And out there in Cardigan Bay are the mythical lands of Cantre'r Gwaelod that illuminates the sea from below with the setting sun, and the utopian dreamworld of Plant Rhys Ddwfn that shines a light from above. However, all these mystical lands are usually hidden by Y Brenin Llwyd, the Grey King, a thick mist that sits on top of Tre'r Ceiri and prevents anyone seeing beyond the ends of their noses. So Môrwen squints and in her imagination she sees a town full of hidden giants, several faraway countries, a submerged land, and a forgotten utopia, and all are equally real in her mythology.

8

Arianrhod and the Menai Strait Monsters

DINAS DINLLE–CAERNARFON–YNYS LLANDDWYN

Arianrhod

Out at sea off Dinas Dinlle is a rock known as Caer Arianrhod, the site of a strange encounter in the fourth branch of *The Mabinogion* that explains how Blodeuwedd, the woman of flowers, transformed into an owl.

Arianrhod was the niece of Math, King of Gwynedd, and sister to lovelorn Gilfaethwy and the malevolent conjuror Gwydion. Her two brothers worked for their uncle Math, who spent his days in his castle at Arfon with his footholder, Goewin ferch Pebin, the virgin who looked after the King's body and rubbed his royal feet. He saw no reason to leave Arfon other than in times of war.

Gilfaethwy was infatuated with Goewin, but she showed no interest in him or his feet. So Gwydion decided to help his brother's love life by creating a war to prise Math away from his footholder. He told his uncle that Pryderi, King of Dyfed, had domesticated pigs, and wasn't it embarrassing that Math, King of Gwynedd, had none? So Math sent Gwydion to Pryderi's court in Narberth, where he enchanted them with stories, for he was the finest storyteller in the land. Then he conjured twelve horses and twelve hounds from toadstools, and persuaded Pryderi to swap them for domesticated pigs. Gwydion herded the pigs home to Gwynedd, but when Pryderi woke from the enchantment, he raised an army and marched north in pursuit of Gwydion. When Math heard of this, he raised his own army, left Goewin in his castle and met the southern invaders at Maentwrog. In the resulting civil war over domesticated pigs, Pryderi was slain.

When Math returned to Arfon, he anticipated Goewin would rub his feet, but she stood tall and explained that while he was away at war, Gwydion watched from afar while Gilfaethwy raped her, 'and in your own bedchamber. Your nephews have shamed you. I am no longer a virgin.'

Math wreaked retribution. He gave Goewin the power to rule Gwynedd by his side on behalf of women. Then he transformed Gwydion into a stag and his brother into a hind and watched as they mated and Gilfaethwy gave birth to a fawn. Then Math conjured them into a boar and a sow until Gilfaethwy gave birth to a piglet, and then he magicked them into wolves till they bore a cub. Math turned the fawn, piglet and wolf cub into three children and his nephews into men, their shame utterly complete.

Gwydion, forever devious, told Math he would need a new virgin footholder, and he recommended his sister Arianrhod. So Math conjured a test of Arianrhod's virginity, but her waters broke and she gave birth to a sturdy boy with thick yellow hair who leapt into the sea. Arianrhod named him Dylan ail Tôn, Dylan of the Waves, the spirit of the water, whose gravestone is a rock off Clynnog Fawr.

Arianrhod, exposed, ran from Math's palace and dropped her placenta, which Gwydion hid in a chest at the foot of his bed. From it, he conjured a boy who he raised as his son. When Arianrhod discovered her brother's deceit, she placed a *tynged* on the boy, that only she, his mother, would name him.

As time passed, the boy asked his name. So Gwydion conjured a ship out of seaweed, disguised himself and the boy as shoemakers, and sailed to Arianrhod's castle at Dinas Dinlle. He offered to measure her feet for a pair of fine shoes, and as she stepped on board, he told the boy to throw a stone, which struck a poor wren between the tendon and bone of its leg.

Arianrhod murmured, '*Lleu Llaw Gyffes.*' (Fair-haired boy with a skilful hand.)

Gwydion revealed himself. 'His mother has named her son Lleu!'

Arianrhod, furious at her brother's trickery, placed another *tynged* on Lleu: that he will only go to war with his mother's permission.

So Gwydion conjured a fleet of invading ships and when Arianrhod saw them approaching her castle, she ordered everyone, including Lleu, to prepare for war. In that moment, the ships vanished, and she realised her brother had tricked her again, so she placed a third *tynged* on her son: he will never have a human wife.

So Gwydion and Math conjured a wife for Lleu from the flowers of oak, meadowsweet and broom, and named her Blodeuwedd, the owl-girl whose story ends the fourth Branch of *The Mabinogion*.

And all the while, Goewin watched the machinations of men at Math's court, and dreamed that the fifth branch would be hers.

Macsen Wledig's Dream

The Roman Emperor Maximus dreamed he was flying over a great walled city at the mouth of a river protected by a large fleet with a gold and silver ship. He sailed the ship over an ocean to a castle with a hall made of gold, silver and precious stones. Two youths dressed in jet black satin were playing chess while a maid sat in an ivory chair wearing an off-the-shoulder white gown held in place with a golden clasp. Emperor Macsen Wledig immediately fell in love, and woke up.

Macsen was consumed by the dream girl, so he searched the land until he found the walled city, boarded the golden ship, arrived in Brython, rode over Snowdon to Caernarfon, entered the great castle, saw the boys playing chess, and met the dream maiden. She was Helen Luyddawc, of the Roman road named after her, Sarn Helen.

Strange Things in Menai Strait

Môrwen wanders along Castle Street, wondering how to cross the Menai Strait to Ynys Môn. The Foel ferry closed in 1954 after a dramatic drop in passenger traffic caused by an increase in cars crossing the suspension bridge. Caernarfon railway station closed in the early 1970s for the same reason, but this is Calan Gaeaf, so Môrwen steps back and catches the little paddle steamer *Menna* to Ynys Môn.

Strange things have been seen in the Menai Strait. In October 1805, a huge worm-like serpent slithered through the tiller hole of the *Robert Ellis* and coiled up on the deck. The crew pushed it overboard with oars, but it chased them until the wind blew the ship to safety. However, the strangest thing was a cat.

The Monster Cat

Henwen was a giant pig, so big the people in Cornwall thought she was about to give birth to something evil, so they chased her into the sea. She swam to Wales and when she passed through Maes Gwenith in Gwent, she gave birth to a tiny grain of wheat and a small bee. At Llonion in Pembroke she bore an even tinier grain of barley and a smaller bee. In Llŷn, she bore a grain of rye so small you'd need a microscope to see it. Then she gave birth to a huge wolf cub and a massive eagle at the Hill of Cyferthwch in Eryri, and when she reached Llanfair in Arfon she bore a kitten that grew into a giant monster cat. As it swam across the Menai Strait to Ynys Môn, the sons of Palug pulled it from the water, but they must have wished they hadn't. Palug's Cat tore 180 warriors to shreds with its sharp claws, and became one of the three monstrous plagues of Ynys Môn.

Sailors in Wales believe the mewing of a cat on board means a bad voyage. If the ship's cat often wipes its face with its paw, trouble is ahead. And if the cat scratches the mast, nothing in the world can save the crew.

Druids

The Roman cavalry marched along Sarn Helen to the Menai Strait and crossed over to Ynys Môn, where they were greeted by wild haired-women and men with painted faces and bodies, dressed in death-black robes and waving fire torches as they lifted their hands to heaven in supplication. Field names bear witness to a bloody battle: *Maes Hir Gad*, Field of long battle; *Cae Oer Waedd*, Field of bitter lamentation; *Bryn y Beddau*, Hill of Graves.

Cadi the Baby Farmer

Safely on Ynys Môn, having avoided the monster cat and the Roman army, Môrwen meets Mr Emerson, a documentary photographer from Norfolk in the early 1890s. He is visiting Menai Bridge to photograph people in their natural landscape and is recording oral stories to complement his images. He buys Môrwen a beer from the Mermaid Tavern and tells her this tale:

'Cadi was a tall red-haired girl from Ynys Môn, who was visited by the fairies. They crept through her keyhole at night, made a noise like the wind, danced around her room, and left money by the fireplace. Cadi had never seen them, and felt she never should.

'Soon she saved enough money to have a baby. She chose a suitably tall man, and they had a child who grew like her parents. One evening at the fair, Cadi's man asked how she earned her living, so she told him about the fairies, the keyhole and the money. That evening she tucked her baby snug in his cradle, kissed her and went to bed. In the morning, her baby and all her money were gone, and in the cradle was a wrinkled fairy child. You see, you should never tell anyone about fairy money.

'As time passed, Cadi became a baby farmer. She had so many babies, she couldn't count them all or remember their names. One day, she went to the woods to gather sticks for the fire, and found a piece of gold. She knew it was fairy money so she never spoke of it. Instead, she bought fine clothes for her children, though her firstborn fairy child always walked around in rags.

'One day, a man knocked at her door, and told her she might be eligible for poor relief for such a large family. She closed the door and hid her well-dressed children in the crog-loft with her gold. She opened the door with her ragged fairy child, and told the man she needed money for she had to dress her precious babies in rags. After the man left, Cadi went to the crog-loft, but her children had vanished, taken by the fairies, and her money had turned into cockleshells from the Menal Strait.'

Dwynwen, Patron Saint of Lovers

Môrwen calls at Plas Newydd to see Rex Whistler's trompe l'oeil, a wall-sized mural painted in 1936 of the view from the house across Menai Strait, though it also included a Venetian gondola, a washing line hanging from a window at Windsor Castle, the painter's stubbed out cigarette, and Neptune's wet footprints on the steps leading from the water.

Môrwen follows the footprints into the painting and finds herself on the beach at Ynys Llanddwyn, the island home of Dwynwen, Patron Saint of Welsh lovers. Dwynwen sits in the ruins of her church and tells Môrwen her story:

'I was one of twenty-four daughters of the fifth-century Christian King Brychan Brycheiniog. I never knew my mother, and father only taught us about compassion and piety, but not about men. He arranged for me to marry a rich man but I was infatuated with a poor boy from the village, Maelon Daffodrill. I ran into the forest, threw off my clothes and cried myself into a dream where an angel gave me two cups. I drank from the blue cup and there was Maelon staring with sad puppy eyes beneath his dark furrowed monobrow. He removed his shirt and was untying his trousers when I cooled, and my cold stare froze the blood in his veins and he turned into a block of ice. Like many women, I blamed myself.

'So I walked for 3 days, then climbed the hill to a pond where three wish-fish lived. I drank from the red cup and was offered three wishes.

'Wish one, defrost Maelon and warm his blood.

'I watched as the ice melted and life returned to Maelon's naked body. I waited for him to fall into my arms but he ran away without a word with his trousers flapping round his ankles.

'Wish two, no man will come closer to me than an arm's length and a palm's width.

'Wish three, everyone must have the chance of finding true love in their own way.

'Then I fled to Ireland with my sister Ceinwen to forget my troubles, and when I eventually returned to Wales, I settled on Ynys Llanddwyn to live in celibacy and solitude. Women gathered 'round me, a commune grew, and then a convent. After I became a spirit, I was made a saint of true, albeit doomed love.

'Pilgrims visited and left offerings in my shrine, drank water from the well, spread their handkerchiefs on the surface of Crochan Dwynwen and cut their names on the rock to see the future.

'Dafydd ap Gwilym wrote poems about me and made a pilgrimage to ask for help to win the heart of his beloved Morfudd, a married woman. Richard Kyffin, rector of Llanddwyn and Dean of Bangor Cathedral, used the pilgrims' offerings to build himself a fine home on the island. As time passed, my church was destroyed by storms, and I was forgotten.

'In the late 1700s, my story was written down by Iolo Morganwg, although it was Vera Williams, a student at Bangor University who changed everything. In the 1960s, she printed four cards designed by Elis Gwyn Jones, and persuaded Y Lolfa to publish them so young lovers could send them to each other on January 25th, St Dwynwen's Day.'

Môrwen walks North past Malltraeth Marsh, where she sees Mr Tunnicliffe painting the wintering swans for his book, Shorelands Winter Diary, a paean of love to wild birds. His work is now in Oriel Môn in Llangefni.

9

Riding a Mammoth Round Ynys Môn

CAERGYBI/HOLYHEAD–MOELFRE–
PORTHAETHWY/MENAI BRIDGE

St Ffraid

A Hawk T.2 jet from RAF Valley roars over Môrwen's head as she crosses Four Mile Bridge on to Holy Island. She walks to the dunes at Towyn y Capel, where St Ffraid was venerated in a wooden church on a mound that contained 400 graves until it was taken by the sea. In Ireland, Ffraid is Bridget, one of the Tuatha Dé Danann, who rode to Wales on a piece of turf after fighting arranged marriages and turning muddy water into beer.

Myfanwy the Holyhead Mammoth

In 1864 the jawbone of Myfanwy the Woolly Mammoth was unearthed during the construction of Holyhead harbour. The jawbone was donated to the Natural History Museum in London, then returned to Ynys Môn, and since then Myfanwy has waited patiently outside the Holyhead Maritime Museum for it to be returned to her. She is a 30,000-year-old hairy mountain of quiet indignation.

There were no mammoths in the Mesolithic, other than in Nan's stories, so Môrwen strokes Myfanwy's trunk and kisses her cheek. They wander past the Stena Line ferry terminal near the grave of Irish chieftain Sirigi, whose death in battle led to Welsh rule on Ynys Môn.

Môrwen follows the map on her phone to the bank of the Alaw, where Branwen is sitting on her own gravestone. 'Honey! Love your hairy friend!'

'She can't speak. She has no jaw,' explains Môrwen, 'Is this where you're buried?'

'I died of a broken heart after the wars between my two countries. I think my bones are under this stone, but I'm not bothered 'cos my spirit lives on in my story. Hey, you can take the girl out of *The Mabinogion*, but you can't take the Mab out of the girl. See you in Gwales,' and Branwen blows a kiss and vanishes, leaving a swarm of black bees to guard her grave.

Môrwen places Branwen's phone on the stone. Her sketchbook smells nicer.

Bonesetters

Following a shipwreck around 1744, two 8-year-old boys who spoke an Eastern European language were found wandering along a beach near Llanfairynghornwy. One died shortly after, but the other was adopted by a local doctor and given the name Evan Thomas. Evan learned the art of the bonesetter, using touch to feel where bones were broken on injured birds and animals, manipulation to knit fractures, and splints to keep limbs still until they mended. Evan passed on his skills to his own son, grandson and great grandson Hugh Owen Thomas, who invented the Thomas Collar to treat osteo-articular tuberculosis, the Thomas wrench for reducing dislocations, and the Thomas splint for mending fractured femurs. Hugh's nephew Robert was the first to use X-rays to diagnose bone damage, and so orthopaedic surgery began with strange things on a beach on Ynys Môn.

St Patrick

On the north coast of Ynys Môn, another Irish Saint was shipwrecked on Middle Mouse during a mission from Pope Celestine to introduce Catholicism to the locals. Patrick had already transformed unbelievers into mermaids and converted a merman to Christianity, and was so sure of himself he left his name everywhere. He swam ashore in Patrick's Cove, drank at Patrick's Well, lived in Patrick's Cave and called the island Ynys Badrig.

Patrick's Church at Cemaes was inherited in 1869 by Henry, third Lord Stanley of Alderley in Cheshire, along with much of Ynys Môn. Henry was a travel writer and vocal critic of British colonialism, who spoke Persian, Turkish and Arabic, and swore he would never marry an English woman. While in Istanbul in 1862, he converted to Islam after marrying Serafina Fernandez Funes of Alcandete, who it later transpired was already married, so they married twice more when they returned to Britain. Henry took his seat as the first Muslim in the House of Lords under the name Abdul Rahman and in 1884 he transformed St Patrick's church into a mosque, with deep blue tiling, geometric-patterned stained glass windows and Arabic iconography.

It's unlikely Patrick ever thought he would help introduce Islam to Wales, or that Abdul knew his beautiful mosque would be overshadowed by Wylfa Nuclear Power Station, nor that the Dalai Lama would once sit there and declare it 'the most peaceful place on earth'.

Cave of the Blue Horse
In the late 1700s, following a family argument, a young man rode over the cliff at Cemaes Bay and his dappled blue-grey horse swam ashore at *Ogof y March Glas*, the Cave of the Blue Horse.

The Mapmaker
Môrwen and Myfanwy wander by Afon Goch, *the Red River*, where a boy from Pentre Einianell farm is drawing the coastline in a sketchbook. Lewis Morris will grow to be an antiquary, literary scholar, land surveyor, book publisher, botanist and self-taught hydrographer. In 1729 he became a customs officer for Holyhead, where he heard the sea captains grumbling about so many deaths within the fishing and sailing communities caused by poor maps and sea charts. With little support from the Admiralty, Lewis surveyed Beaumaris in 1737 and by the end of the year he had produced a folio of eleven detailed charts of the Welsh coast and seas from Llandudno to Milford Haven. His complete maps were published in 1748, and updated regularly by his son William. The Morris maps saved many ships and lives at sea. The *Royal Charter*, unfortunately, was not one.

The Moelfre Heroes

The *Royal Charter* was a fast, steam-driven sailing clipper, built in Sandycroft on the Dee estuary in 1855. She had fifty first-class berths and twenty-eight state rooms, and had already sailed to Australia via Cape Horn in a record sixty days. Captained by Thomas Taylor, she left Melbourne on 26th August 1859 en route to Liverpool, with 118 crew and 371 passengers, many of them miners who had emigrated to make their fortunes in the Australian Gold Rush.

At around ten o'clock on the night of 25th October, she sailed into a storm north of Point Lynas. The anchors and masts snapped, the engines were switched on, and Captain Taylor ordered all passengers to the lower decks for safety, shortly before he was killed by a falling lifeboat. The ship broke in two on the rocks between Red Wharf Bay and Traeth Lligwy, cracking open the hull and washing those below into the sea.

Able Seaman Joseph Rodgers swam ashore with a rope around his waist to rig up a line between the wreck and the beach with a makeshift bosun's chair. People from Moelfre joined the line and helped rescue twenty-one passengers and eighteen crew, including two young boys tied to planks by their father. More than four hundred people drowned and their bodies were washed up on the shore.

The newspapers accused Captain Taylor of incompetence and drunkenness, and claimed the passengers perished because they were weighed down by the gold in their pockets, which the local people stole from the dying in exchange for help. *The Daily Telegraph* called for the death penalty for the, 'greedy Cambro-British thieves'. In response, the 'Moelfre Twenty-eight' wrote to *The Times* to condemn the London journalists, whose lack of Welsh meant they failed to understand that the majority of the gold was handed in to the authorities.

Stephen Roose Hughes, rector of St Gallgo's Church, ministered to those caught in the wreckage, shared their last moments and laid out the bodies in his church for identification. He carefully recorded birthmarks, rings, tattoos, eye colour, then salted and disinfected the bodies, buried them with dignity and listened to every survivor's story. He praised the local fishing community for the lives they saved, and consoled them as they paid the costs out of their own pockets.

Hughes had seen Hell, and was doing all he could to help the souls of the drowned from going there. He and his wife wrote to their relatives, and received more than a thousand letters of thanks, to which they replied. He criticised the government for caring more about gold than people, and was consequently threatened with arrest if he did not stop.

Charles Dickens visited Moelfre and wrote, *The Uncommercial Traveller*, which supported Hughes and condemned the press and government. They kept in touch and exchanged gifts, until on 4th February 1862, Hughes died aged 47, worn out by emotion and exhaustion. The local people never forgot him, and a service is held every year at St Gallgo's Church to celebrate this true Welsh folk hero.

The other hero, Able Seaman Rodgers, real name Giuseppe Ruggier from Malta, is commemorated in a bronze sculpture at the Moelfre Seawatch Centre. He died in 1897 aged 68 and was buried in a pauper's grave.

The Snakeskin Charm

In the early 1800s, a Moelfre fisherman saw a strange woman dressed in a flowing white gown struggling in the sea near Lligwy, close to where the *Royal Charter* was later to sink. He waded into the waves, pulled her ashore and helped her climb the hill, where she sat on the capstone of the Lligwy Cromlech. She had been thrown off a ship by men who called her a witch, and she told him if she showed her true likeness the fisherman wouldn't be staring at her like that. To thank him, she gave him a rolled up snakeskin as a charm to keep him safe at sea, but said he must keep it hidden and handle it only once a year. Then she ran to the beach, leapt over the rocks and swam away.

The fisherman buried the snakeskin by the cromlech, and from that moment he enjoyed good fortune. Every year he dug up the charm and held it to his cheek before burying it again, until one day he caught no fish. When he went to the cromlech, the charm had disappeared.

On his deathbed, a neighbour admitted stealing the charm and returned it to the fisherman along with his good fortune. He gave it to his eldest son, who left Moelfre for a new life in Australia. The son made a fortune in the gold rush before the charm vanished. Who knows who has it now?

The Llanddona Witches

One stormy night, a boat came ashore in the shadow of Castle Rock in Red Wharf Bay containing a cargo of ragged men and women, but no rudder or oars. The locals thought they must be escaped criminals set adrift to drown as punishment, so they pushed the boat out to sea, but were forced back by a spring of fresh drinking water that gushed like a fountain from the sand.

So the strangers made a home on the commons at the edge of the village. The men became smugglers who could remove their red neckties and release black flies into the eyes of prying customs officers who failed to see the rum and brandy in front of their noses.

The women were small with red hair. Siani Bwt was 44in high, had two thumbs on her left hand, and told fortunes once a week in a rented room in Caernarfon, where people flocked to see her. Big Bela, Lisi Blac and Elen Dal lived by begging, and if anyone refused to give them milk or potatoes, they turned into gnarled hares and cast spells. Big Bela once cursed a man called Goronwy Tudor when she was in the form of a hare, so he moulded some witch's butter from decayed trees into the witch's shape and stuck it with pins, which caused her to scream and turn back into a woman until she released him from the curse.

Building Bridges

Nestled between the two bridges over the Menai Strait is Ynys Gorad Goch, where people gathered cockles, mussels and oysters, emptied the fish traps and rested while they herded cattle through the Swellies. The Britannia Bridge, the world's first box-girder bridge, was rebuilt after it burned down in 1970 when teenagers set fire to the wooden railway sleepers with a piece of burning paper they were using as a torch. The bridge's four lions survived, which inspired local poet and seafood seller John Evans, *Y Bardd Cocos*, the Cockle Bard, to write:

Pedwar llew tew
Heb ddim blew:
Dau'r ochr yma
A dau'r ochr drew.

Four fat lions,
without any fur:
two this side,
two on the other.

In 2023 Welsh singer-songwriter Ren Gill from Menai Bridge raised £21,000 for Beaumaris lifeboat crew after they searched for his friend Joe Hughes, who had vanished from the suspension bridge. Ren has spent his life battling his own debilitating illness before becoming number one in the album charts in 2023, all by word of mouth, busking and videos, yet he never forgot the RNLI.

Halfway across the bridge, Myfanwy's trunk strokes Môrwen's shoulder and she tramps back to Holyhead to see if anyone has returned her jawbone.

10

Down the Rabbit Hole With a Goat

BANGOR–CONWY–LLANDUDNO

Penrhyn

On the south side of Menai Strait, at the mouth of the Cegin river, stands Porth Penrhyn, its fading industrial past a reminder of the slate and slave trades. Built on the backs of the quarry workers at Bethesda and the enslaved people on plantations in Pennant, Jamaica, and named after the family who lived for generations at Castell Penrhyn.

On 31st August 1844, quarryman Hugh Hughes was fishing near the cast-iron bridge when three gamekeepers accused him of poaching and trespassing on Pennant family land. Hugh explained his family had fished there for generations. A fish trap on the Bangor Flats goes back to 1556, and there is a wicker fishing weir near Lord Penrhyn's bathhouse at the end of his private jetty. So Huw took Pennant to court and won £5 damages and the right of working people to fish. The Welsh Coast Path, however, has no such permission.

A Welsh Romany Girl

Eldra Roberts spent her childhood years at Penrhyn in the 1920s, where her father Ernest worked as one of the third Baron's river-keepers. As a Welsh Roma girl, the forest was her playground where she learned to tickle trout, keep ferrets and net rabbits, while her grandfather Reuben taught her the triple harp, and her mother Edith told her wonder tales, for she was descended from the renowned Wood family storytellers. Her life on the estate was documented in *Eldra*, a Welsh language film made by a Romany production company, in which Eldra tames a wild fox long before her people were forced to quit their travelling culture.

The Women of Penmaenmawr

On the hills above Penmaenmawr are stone circles. One Sunday, three women climbed the hill to winnow corn and dance and sing, but God disapproved of work and pleasure on the Sabbath so he turned them into three stones the colours of their dresses: one red, one blue, the other white. The colours have faded now, but the stones still dance when the wind blows.

Below the hills, a causeway crossed Traeth Lafan from Penmaenmawr to Puffin Island until the sea washed all the houses away and left only mudflats and stranded ships.

Violet's Leap

In January 1909, an expensive Minerva car was found hanging over the sea wall at Penmaenbach. The windscreen was smashed and the driver, 24 year-old Violet Charlesworth of Rhyl, presumed drowned, until a police search discovered she was on the run dressed in a red cloak. She had countless creditors, having borrowed money from rich gentlemen on the promise she would pay it back with interest or marriage when she received an inheritance from her godfather on her twenty-fifth birthday. She was tracked down to Oban, where she was living under the name Margaret MacLeod. After her release from Aylesbury jail, she vanished again, leaving Violet's Leap as a lasting memory of the Welsh fake heiress.

Rogark Dolls

Throughout the 1950s, 100 women worked at home in Penmaenmawr for Rogark dolls. They assembled plastic body parts, dressed them in national costumes, added eyes and wigs, and delivered the finished dolls by hand to the company office at Penholm, a large house on Bangor Road, to be boxed and sent to the distributor in Liverpool. Tourists bought them from souvenir shops, little knowing these cute national stereotypes like 'Gwyneth from Wales' were helping the women of Penmaenmawr earn a living.

The Mermaid, the Saint and the Witch

The sound of the tide in the Conwy estuary is known as 'Dylan's Death Groan' after Arianrhod's son, the water spirit. A mermaid died here after fishermen refused to release her when she was caught in their net. As she dipped her tail in the water for the final time, she cursed the people of Conwy to be as poor as the fishergirls who gathered mussels on the estuary in the hope of finding rare pearls. The girls sold their pearls on the other side of the river in wealthier Llansanffraid, where the church is dedicated to St Ffraid, the Welsh St Bridget whose support of women is legendary. In 1594, Ffraid must have looked on in disbelief when her church was used for the witchcraft trial of Gwen Ferch Elis of Wrexham, who was executed in Denbigh.

Afanc

In the Conwy river lived a monstrous *Afanc*, a wild hairy creature with a scaly tail and huge yellow teeth, who tore down trees, built dams, destroyed crops and flooded farmland.

The people sent for Huw Gadarn, Huw the Mighty, who arrived in Conwy riding his plough pulled by two long-horned oxen, the Ychen Bannog. Huw considered himself to be the first farmer. He had turned wild boar into pigs, jungle fowl into chickens and longhorn cattle into Welsh Blacks, so he could domesticate a monstrous hairy creature.

Huw asked for a volunteer to be bait to trap the *Afanc*.

'You!' said Huw, pointing at Môrwen, who was sitting beneath a crack willow on the riverbank opening mussels to make a pearl necklace.

'I don't want to be bait.'

'Don't worry,' said Huw, 'I will leap out and slay the wild hairy monster.'

'I'm not worried. The *Afanc* isn't a monster. It's cute.'

Huw watched until twilight, when the river bubbled and out crawled the *Afanc*, dripping with pondweed, thrashing its scaly tail and gnashing its yellow teeth. Môrwen held out her arm, it sniffed her hand and laid its head in her lap.

'No one's going to hurt you,' she whispered.

Huw leapt out from behind a tree, chained the *Afanc* to his plough and the Ychen Bannog dragged it away. The *Afanc* whimpered and clung to Môrwen's breast, but the cattle hauled it through Bwlch Rhiw'r Ychen, the Oxens' Pass, where an eyeball popped out and formed Pwll Llygad Ych, Pool of the Ox's Eye.

Huw looked into the *Afanc*'s watery eyes and his heart melted. That annoying girl was right. It wasn't a monster. It was cute. And it didn't need to be domesticated. So he released it into Llyn Glaslyn and told no one he hadn't slain it.

Afanc means 'beaver' in Welsh.

The Smallest House

Môrwen knocks on the door of the Smallest House in Conwy. Quay House is less than 2m wide with one room downstairs and a ladder to the upstairs room, where there's no toilet. It was home to fisherman Robert Jones, who was 1.9m tall and slept with his feet out the window. When he left in 1900, it became home to the *pobl bach*.

A little woman opens the door and invites Môrwen in, lights a fire, smooths the quilt on the bed, fills a hot water bottle from the tap behind the stairs, and plumps up a duck feather pillow. Môrwen stretches out in luxury until she is rudely woken by the *tylwyth teg* dancing round the room to the sound of a solitary fiddler who plays faster and faster until they all crash in a heap in the middle of the floor, laughing and giggling. The little woman serves oat milk porridge for breakfast and asks only for a cockle, a winkle and a mussel as payment. It's still Calan Gaeaf, and barely a moment has passed.

Llys Helig

Môrwen crosses the pedestrian suspension bridge, creeps round the Llandudno Junction road and railway spaghetti, and climbs up to Deganwy castle, which overlooks the submerged land of Llys Helig.

Old Helig ap Glanawg wanted his daughter to marry a wealthy young man who would manage his castle. She told him she could run the castle alone, but Helig insisted it was a patriarchal world and only a man rich enough to wear a golden torque round his neck could inherit the castle. So she left her stubborn father and went to live in the forest, where she built beehives, sold honey and became one of Conwy's renowned beekeepers.

One day, a young entrepreneur with carefully dishevelled hair applied for the job, but Helig rejected him as he had no golden torque. The young man had no morals, so he killed a rich man, removed the torque from his neck and buried the body in the woods. As he washed the blood and dirt from his hands, a voice told him a curse would fall on his great, great grandchildren. The young narcissist had no thought for the future, so he returned to Llys Helig with his golden torque.

After Helig died in mysterious circumstances, his palace passed to the young man, who treated his servants as slaves and partied during plagues. Time passes quickly in fairy tales and soon his first great, great grandchild was born. An old blind harper was hired to play for the christening in exchange for as much mead as he could drink. The harper knew the story of the curse, so he asked a kind servant girl to warn him if she saw anything weird.

As the end of the evening, the harper sat at the foot of the stairs with the last of the mead, when the servant girl rushed in and told him the cellar was flooded and full of leaping fish. She took his hand and led him up the hill to the safety of Castell Deganwy, while the partygoers were so busy draining suitcases of wine they never noticed the rising floodwaters. One by one they fell asleep, and never woke up.

The Great Orme Dragon

A Viking longship from the Northlands was approaching the Great Orme when the wind blew and a monstrous serpent churned the sea. It lashed the water with its tail, claws ripped the sails, teeth splintered the wooden hull, and the Vikings were thrown into the sea. The survivors swam ashore, and hid in Ogof Cythreuliaid, the Devils' Cave, below Rhiwledyn Cliffs, but the serpent breathed fire into the cave and roasted them alive, then curled up and slept to the sound of screaming. And that's why the Great Orme looks like a fire-breathing dragon silhouetted in the evening sun.

Poor Hannah

Hannah from Glanwydden worked as a beggar girl on the seafront in Llandudno, where she sold elfwort potions made of elecampane. She could never look people in the eye and was bullied for being unworldly, but she never cried or answered back. If anyone was rude to her, she simply painted their face on a peg doll and stuck it with a pin.

One evening, the vicar found her in the graveyard, accused her of witchery and drove her out of town with a rowan stick. The people chased her to Rhiwledyn Cliffs, so she hid in Ogof Cythreuliaid with the burned bones of the Vikings. The people threw flaming torches inside but there were no screams, only an unworldly silence. Hannah was never a girl to cause a fuss – and, unlike the Vikings, she knew her way through the subterranean tunnels of the Bronze Age copper mine beneath the Great Orme.

Alice, the White Rabbit
and the Goats of Llandudno

Môrwen has followed Hannah through the tunnels and caverns beneath the Great Orme, where more than 30,000 animal bones and tools have been found amongst the cave spiders and blind shrimps, including a 13,000-year-old horse's jaw bone decorated with zig-zags. As she stands outside Kendrick's Cave, she smells something stinky. She is surrounded by a herd of weirdly grinning goats with pointy horns.

The Llandudno Goats rule this town. They are descended from wild Markhor, *snake eaters*, from Kashmir. People move aside to let them pass as they escort Môrwen down to the railway station, where the wooden statue of Alice climbs down from her plinth.

'Curiouser and curiouser,' thinks Môrwen as she follows Alice and the goats to the town square, where the White Rabbit consults its pocket watch.

'You're a very scruffy girl,' says Rabbit.

'And you're a very rude wooden rabbit,' says Môrwen.

'Are you ADHD?'

'No, I'm Mesolithic.'

The Wonderland statues have been here since 1933 after someone realised that Alice Liddell and her family spent their summer holidays in a house called Penmorfa on the West Shore, and Lewis Carroll visited at least once. He included local landmarks in *Through the Looking Glass*, like the two offshore rocks known as the Walrus and the Carpenter. Curiouser still, in 2008 a developer was given permission to demolish Alice's house.

Alice, the White Rabbit and the goats take Môrwen to the pier, where she meets Jason Codman and his 170-year-old Punch and Judy Show, the oldest in Wales. As the crocodile appears with a trail of sausages, they fall down a rabbit hole into the forest of 160 wind turbines growing from the sea, which sing 'Salmon-chanted evening' in time to the revolving blades. From the beach the windmills look like Bran's ships approaching Ireland. From the sea they look curiously like pieces on a chessboard.

11

The Dancing Girl and the Headless Hermit

Madoc

Môrwen pops out of the rabbit burrow outside the Harlequin Puppet Theatre in Rhos-on-Sea, where Mr Bimbamboozle is performing amazing magic tricks and Hans Christian Andersen's *Swineherd* is being told with marionettes. A man on the beach is studying an ancient map of the oceans filled with elephant-headed pig-fish, scaly owl-faced serpents, two-tailed mermaids and shipwrecked sailors with silently moving fish mouths. This is a reflection of the year 1171.

'What you doin'?' asks Môrwen.

'I am leaving in search of Utopia, Taliesin's fabled land beyond the looking glass of the sea. I am weary of the civil wars in my country, so I have built a wooden ship nailed together with stag's horn, and named her *Gwennan Gorn. Madoc ap Owain Gwynedd i fi'* ('I am told I live in Cloud Cuckoo Land'). The man, Madoc, offers Môrwen a beer.

'Can you give me a lift to Prestatyn?' Môrwen downs her beer in one gulp.

The *Gwennan Gorn* sails east past Abergele, where a replica gravestone in the churchyard carries the inscription, *'Yma mae'n gorwedd yn monwent Mihangel, gwr oedd a'i annedd dair milltir yn y gogledd'* ('In this churchyard lies a man who lived three miles to the north of it'). To the north is sea, though it was swamp forest when Môrwen played there as a child.

They sail above the 7,000-year-old stumps of birch, alder, elm and hazel on the seabed at Coed Mawr y Rhyl, still visible at low tide off Splash Point. Strange things have been found here. A 5,000-year-old deer antler, a Stone Age axe head made of rock from the quarries at Penmaenmawr, the wreck of the *Resurgam* submarine and a custom-made bungalow built by a tram engineer who misjudged the level of the floodwaters. Elephants from Billy Smart's circus bathed here in the 1960s near West Parade, the birthplace of nightclub manager Ruth Ellis, the last Welsh woman to be hanged for murder.

Môrwen gives Madoc a *ci-corc*, a lucky cork dog made from her beer bottle cork with matchstick legs, ears and a tail. She dives into the sea and swims towards Prestatyn beach, knowing she will never see Madoc again.

Esmeralda the Dancing Girl

A woman with long grey hair dances along Prestatyn beach and holds out her hands to Môrwen, who is suddenly shy.

'I'm too clumsy to dance.'

'My dear, everyone loves dancing.'

'Not if you were dragged to a *twmpath* as a child!'

'When I was young, I danced with the Welsh Romany, then through the theatres of Paris and London, where Dante Gabriel Rossetti painted me as one of his pre-Raphaelite dream girls, and Victor Hugo wrote the part of the gypsy in *The Hunchback of Notre Dame* about me. I had beautiful black hair like yours. I am Esmeralda the Dancing Girl.'

'I have Calan Gaeaf hair. I've just passed through the veil. I'm Môrwen the Mesolithic Girl.'

Esmeralda kisses her. 'Madoc told me you were coming. You look so young, yet you are older than me. I was born in 1854, one of Noah and Delia Lock's fourteen children. We were travellers, played fiddles, sold wicker baskets and traded at horse fairs in North Wales. I was in my teens when my parents married me to Hubert Smith, a 52-year-old solicitor from Bridgnorth, but I was not a bird to be kept in a cage. I flew like Branwen's starling. I hit Hubert with a silver candlestick and told him I'd been enchanted and needed a charm from a *gozvalo gajo*, a conjurer. It was true, but the *gajo* was my lover, a romantic young folklorist, Francis Hindes Groome, who was writing a book of gypsy folk tales. He was married to another traveller, Britannia Lee, but we eloped to Germany to live a bohemian life, where I earned a living on the stage as the Gypsy Dancing Girl.

'When we both divorced, oh it was scandalous. It was all over the papers. We were famous, you see. Francis' father was Archdeacon of Suffolk, who thought his son was crazy running away with a dancing girl, and crazier to think he could earn a living writing a book of folk tales.

'My uncle, the Welsh Romany harper John Roberts of Frolic Street in Y Drenewydd, wrote to Francis, "I beg to say that I have read all about what happened and felt sorry for you and Esmeralda; it is to be hoped that you are both happy now." I loved my uncle John. I gave him extra cake and tobacco at family gatherings, and danced with him. He was so happy.

'I married Francis in 1876, and we moved to Edinburgh, where he worked as an editor on *Chamber's Encyclopaedia*, but our relationship was a storm at sea. When we separated, Francis wrote, "We must never meet again on this side of the grave." His *Gypsy Folk Tales* was published, but he was dead at 50.

'I took to the road in my green and yellow caravan, and settled at Pendre Farm on Gronant Road. As I grew old, I never forgot how to enjoy storms at sea. I was run over by a bus near the Cross Foxes in 1939 and am buried in Rhyl, so Quasimodo's wild gypsy will haunt these beaches for all time.'

The Ring in the Fish

Queen Nest, wife of Maelgwyn, the Floating King of Gwynedd, lost her wedding ring while bathing in a pool where the River Elwy meets the Clwyd River near Rhyl. Her bad-tempered husband threw a tantrum, so the Bishop of St Asaph invited them to dine on fresh trout caught in the river. As Maelgwyn sliced into his fish, Nest's ring fell out.

Around 1870, a woman was gathering shells on the beach, when her ring slipped off and was washed away by a seventh wave. The next day she bought fresh herring to fry for dinner. When she cut open the belly of the fish, what do you think she found? The ring?

No. Guts.

The River Spirit of the Dee

It's raining old ladies with sticks as a flock of Brent geese fly in from the sea to escape the storm that is blowing across the sands at Point of Ayr. Seaweed hangs limp on back doors, a seaman has a pain in his arm, and a woman stands at the water's edge, arms in the air, shrieking into the wind.

She is Aerfen, Dyfrdwy in Welsh, Dee in English – the spirit of the river and creator of tidal bores and storm surges. People and fish have felt the rasp of her tongue, for she demands three lives a year in exchange for those who dare to cross her. A Roman army from Chester once waded over to Hilbre Island and prayed to St Werbergh to part the waters to allow them to invade Wales. Naughty Welsh children have tried to cross the other way in small boats before they are trapped by the onrushing tide.

As Aerfen's war cry echoes round the exposed beach, Môrwen crouches behind the rusting Talacre lighthouse where Raymond the keeper, dressed in a sailor's cap and uniform, is forever doomed to stare out to sea after he died of a fever induced by lovesickness. It's teeth-achingly cold.

A ragged man appears. 'May I join you? I am rather scared of the river spirit. I am the Headless Hermit of Talacre.'

'You have a head, though?' says Môrwen.

'I'm speaking metaphorically. I'm homeless. A witch in Prestatyn cursed my ancestors to lose our heads for selling Talacre. One died in a light aircraft crash, another was trampled by a crazy horse in Kenya, a third lost a fist fight, and I am the current lord, living in a driftwood hut on the beach.'

'Does your head unscrew?' Môrwen reaches out her arms.

The Hermit wraps his scarf tightly around his neck.

A natterjack toad snuggles into Môrwen's skin coat to escape the wind. This rarest of Welsh amphibians has its own Natterjack Walk at Raymond's lighthouse. The Wise Old Toad of Cors Fochno would be green with envy if he knew (which he probably does).

A Kemp's Ridley sea turtle hauls up the beach and rests its hooked beak in Môrwen's lap. This even rarer reptile has travelled 4,000 miles from the Gulf of Mexico to eat jellyfish.

Suddenly, Aerfen lowers her arms, the wind drops and the river spirit melts into the water. The Headless Hermit, the toad and the turtle return to their homes, as Môrwen follows the industrial estuary through the desolate remains of Point of Ayr colliery. Spirits and ghosts walk the rail lines, memories of 150 miners laid off in 1909 when a fire closed the pit for fear of flooding. There are graves beneath the estuary in the two underwater mineshafts.

On 26th February 1990, the sea wall gave way at Ffynnongroyw and twenty homes were evacuated. Fishermen Joseph Maelor and John Roberts spent the night in their boat *Little Marvel* on Hilbre Island and returned in the morning with a record catch.

The Fun Ship

The *Duke of Lancaster* was launched in 1955 as a British Railways ferry and turned into an amusement arcade in 1979, and ever since then it has been gently rusting near Mostyn Docks. It was spray-painted by the Latvian graffiti artist Kiwie, with the help of British-based Snub23, Spacehop, Dan Kitchener and Dale Grimshaw. One artwork portrays the ship's first captain, Jack Irwin. Môrwen is given a spray can, a paint-stained hoodie and a pair of frayed cargo pants and writes her name in big friendly letters.

The Honest Man

The estuary turns from barren sands to urban streets, so Môrwen calls into the Llety Hotel in Mostyn to meet the antiquarian and story-gatherer Thomas Pennant. In the late 1700s the hotel was named The Honest Man, and was owned by Mr Pennant's grandfather, who held parties in a treehouse in the pub garden, where he drank beer brewed by landlady Jane o Llety. The cellar was stocked with port wine smuggled in through a tunnel connected to the beach, and when the customs men came, a group of miners loaded the bottles on to carriages and hid them where no one would ever find them.

The Fairy Oak

Mr Pennant tells Môrwen the story of the old stag's horn oak tree known as the Fairy Oak that grew near his house at Downing. A poor couple with a 'peevish' baby thought the child was a changeling left by the *tylwyth teg*, so they placed it in a cradle beneath the Fairy Oak overnight. In the morning their baby was sleeping peacefully and the creature was gone, proving without doubt that changelings really do exist.

Gwenffrwd's Head

Mr Pennant escorts Môrwen to the Greenfield Valley, where the hill is made of spoil waste from the underwater mineshafts at Bettisfield Colliery. He introduces her to Gwenffrwd, St Winifred, whose shrine is here and whose head is stitched to her body.

'You look like the Bride of Frankenstein!' says Môrwen.

'Dear rude child,' says Gwenffrwd with a beatific smile, 'let me explain. A brutal young man called Caradog Sychnant fell in love with me, so I hid in the forest here in the Greenfield Valley. Caradog followed me and told me to remove my gown, as it was his right as a man. I explained it was my right as a woman to refuse, but he grabbed my hair, raised his sword and cut off my head. I rolled down the hill and landed near the church where a fresh spring flowed. My great uncle St Beuno found me and stitched my head back on my body, then conjured the earth to open up and swallow Caradog. That's why my home is known as Holywell, and why I have stitches round my neck.'

'There is dark history behind our myths,' Mr Pennant chips in. 'Gwenffrwd's spring provided power for the copper industry to make black manillas, ornaments worn around the legs, arms and neck, and used as money to buy and sell African people.'

'Can I get to the Severn Estuary from here?' asks Môrwen, who hasn't heard a word.

Mr Pennant answers, 'Certainly. Follow Wat's Dyke South from Basingwerk Abbey, where a monk once followed the song of a nightingale, and when he returned the abbey had crumbled to dust. Or catch the train from Chester to Bristol, but you would be on the wrong side of the border. So why not step through the veil between worlds? It's still Calan Gaeaf, after all.'

Mr Pennant gives Môrwen a bottle of Jane o Llety's beer, which she drinks in one gulp, closes her eyes, curses three times, imagines she is Dorothy from Kansas, and steps through the veil again.

12

Sabrina and the Salmon Children

CASGWENT/CHEPSTOW–HAFREN/
SEVERN–CASNEWYDD/NEWPORT

Sabrina

Frozen to the bone, teeth chattering, hair windblown, Môrwen stands in a cathedral of white pillars beneath the M48, where the river sisters Severn and Wye wait to meet their lover the sea. She has travelled through the veil from Holywell on the Dee Estuary to the Severn Bridge near Chepstow, over 200km in the blink of a crow's eye.

The thundering cars and lorries overhead are about to make her scream when a girl appears, wraps a duvet around her shoulders, and rubs her vigorously, just as Myra did when she first arrived at Llanina.

'Aw, that fe-e-els so-o-o-o-o good,' the blood warms in Môrwen's body and she is a child again.

'I've been expecting you, honey.' This girl is another water spirit, more human than Aerfen, and she has many names. Hafren in Welsh, Severn in English, Sabrina in myth and literature.

Sabrina pours hot water from a thermos into a mug and adds a little something from a hip flask. Steam fills the air as Môrwen warms her cold hands on the mug and downs the contents in one gulp.

'Waw! What's that?'

Sabrina refills the mug, 'Juniper berries and seaweed. A Turkish fairy who lived in a magic lamp gave me the recipe. She called it *djinn*.'

The two girls share the *djinn*, and giggle as Sabrina tells her story.

'I'm the illegitimate daughter of Locrinus, King of Loegria in the English midlands 3,000 years ago. My mother was Estrildis, daughter of Humber the Hun, who my father drowned in the river that now bears his name. Locrinus and Estrildis became lovers and I was born.

'I didn't know my father was already married and had a son with Gwendolen of Cornwall, who threatened to split open his head with a battle axe if he ever saw Estrildis again.

'So I grew up alone in my mother's castle at Abermule by the River Severn. I grew to dislike my absent father and I told him he was like a cruel river that flooded fields, killed crops, and starved the people who wanted to love him.

'When my stepmother Gwendolen discovered my father was still seeing my mother, in her fury she raised an army in Cornwall, slew Locrinus in battle, pronounced herself queen with her son Madan as heir, and ordered Estrildis and her daughter be drowned in the Severn estuary.

'But I didn't drown. I became the spirit of the river that separates Wales from her noisy neighbours. However, I can't forget the cruelty of my father and stepmother, and on 30th January 1607 my memories overwhelmed me and I caused a tidal bore on the Severn that flooded the land and drowned 20,000 people.'

The two girls have finished the *djinn*.

'The car noise hurts my ears,' says Môrwen.

'We'll go somewhere quieter.' And the River Spirit and Mesolithic Girl walk arm in arm across the water to Treacle Island, also known as St Twrog's Island, St Tecla's Island or Chapel Rock. It's only accessible at low tide, and spooky when the mist drifts in and no one can see you. Môrwen feels invisible to those roaring across the bridge. For a moment, there is silence.

The Severn Ferry

The *Severn Princess* was a car ferry that travelled across the estuary from the banks of the Wye to Aust on the English side. Shortly before it closed in 1966, Bob Dylan – who borrowed his name from Dylan the Welsh poet – was photographed at the ferry terminal on his way to play at the Capitol Theatre in Cardiff, following a miserable night in Bristol during the infamous 'Judas' tour. In Cardiff, life improved when Johnny Cash joined him for an impromptu post-gig session and Angela Carter wrote an understanding review for *The Observer*, noting that, 'Huck Finn ... has grown up into the first ever all-electronic, all existential rock'n'roll singer.'

In 2005, Dylan used the iconic ferry photo as the cover of the DVD release of the documentary film *No Direction Home*, shortly after the *Severn Princess* was found dilapidated on the west coast of Ireland and returned to her home in Chepstow.

A Supernatural Well

A well at Mathern was believed to be fed by spring tides from the Severn. To test the theory, a man took a log from the well and dropped it in the open sea. Four days later, it was back in the well, handy for people to stand on while they washed. The man suspected the supernatural, so he buried the log in the ground, and four days later it was back in the well again. He died within the year.

The Vanishing Village

The Severn railway tunnel begins at Sudbrook, once a thriving fifteenth-century town before it vanished, only to be rebuilt in the 1870s to house the labourers who lived in small terraces on one side of the street, while the engineers and managers lived in large bay-windowed houses on the other. Construction took thirteen years and when it opened in 1886, it was the longest main-line railway tunnel in the UK.

Black Rock Lave Net Fishermen

At the water's edge near Portskewett, the Black Rock Lave Net Fishermen catch salmon with hand-held conical nets woven from pollarded willow, and larger putcher nets fixed to a long fence of staves. They wait till the high tide covers the nets, and if they see movement in the shallows, they sometimes run through the water to spear fish or catch them by hand. There are some 400 nets out there, and it takes half a day after each tide to check them.

The salmon are sent to the Severn and Wye Smokery and then by lorry to swanky London hotels and restaurants like the Savoy and the Ritz, where customers slice into their fish unaware of the unique lives of the Black Rock Lave Net Fishermen. They still teach their children how to weave nets, navigate the dangerous estuarine gulleys, diversify into farming to earn a living, and clean up the increasing rubbish that washes up on the beach. The tradition of the Severn fisher-families is to be cherished.

She-Wolf

Môrwen and Sabrina are sleeping off the after-effects of the *djinn* at Black Rock, when they are woken by a cold nose and wet raspberry tongue. A She-wolf lowers her tail, paws the ground and rolls on her back in submission to the Mesolithic Girl who was a sister to the wolf packs that lived here before they were hunted to extinction in the twelfth century, around the time Llywelyn Fawr's dog Gelert famously killed one in Beddgelert.

She-wolf allows Môrwen to ride on her back as she lopes along the shore, escorted by a family of wild boar who were reintroduced in the Forest of Dean in the 1980s. Môrwen feels like Princess Mononoke San, guardian of the forest.

Sabrina watches them go. She must clean up the plastic bottles, twisted metal and invisible liquids from her shoreline, and keep an eye on those nuclear power plants on the far bank.

The Caldicot Re-enactment

At Castell Cil-y-coed, Caldicot Castle, a strange event is taking place. A group of German Second World War soldiers have gathered outside the main entrance, surrounded by vehicles draped in camouflage nets. An officer strides toward Môrwen in squeaky boots.

'Tidy horse you got there, love. Which army you from?' He speaks in a strong Cardiff accent.

'I don't have an army.'

'Ah, you're Scum? Through the gate and two fields along.'

'I'm not scum. I'm Mesolithic.'

'Don't think we have a group that old. Scum are ninth and tenth centuries. You've got wicked hair, though, you'll fit in a treat, if you don't mind dyin'.'

Môrwen is about to give him a mouthful when a detachment from the Spanish Civil War International Brigade approaches. There is much gesticulating, pushing and shoving, until the Brigade walk away muttering something about bloody Nazis always getting the best spots.

Boat Building

The remains of a thirteenth-century ship was found buried in the mud 500m from the sea wall at Magor Pill in 1994. When the ship sank, this area was part of Abergwaitha, a now submerged port that served the monks at Tintern Abbey and the towns of Newport, Caerleon, Chepstow, Monmouth and Hereford. Further along the estuary at Barlands Farm, a Roman flat-bottomed boat was found. It had sails and oars and travelled up the tidal creeks and along the beaches, loading and unloading on to wooden jetties.

Mesolithic Goldcliff

Each autumn, Mesolithic people lived on an island covered in oak forest at Goldcliff, where they cooked deer, wild cattle, pigs and eels with hot stones, built wooden boats with bone and flint tools, and caught fish in traps in the estuary. Môrwen stands in one of the sixty footprints preserved in the peat. Some were made by birds and deer, but most are children. They could be hers. The children chased each other, swerved to prevent being tagged, slid down the banks, and splayed their toes to stop falling into the water. Nan treated their cuts and bruises with herbs and gave them nasty-tasting potions for worms and diarrhoea. Not all kids grew up.

Newport Protests

She-wolf and Môrwen cross the river in the gondola hung beneath the transporter bridge. They stroll down Commercial Street past the Westgate Hotel, unseen by passers-by. In 1839, almost 100 armed soldiers were stationed here, waiting for 10,000 Chartists who were marching on Newport in support of democracy, universal suffrage and the rights of non-property owners to vote. There was bloodshed and lives were lost.

Eighty years later, on 6th June 1919, there were race riots in Newport, when homes and restaurants belonging to African, Chinese and Greek people were attacked. Tom Savage, fireman and merchant seaman, was born Kru in Sierra Leone and came to Newport with his brothers in 1917 to work on the ships. He lived in the African Boarding House on George Street with dockworkers, ship's cooks and Merchant Navy men. He died in 1940, one of many Newport sailors, when SS *Cheldale* collided with a ship in Durban.

She-wolf leaps out of the way of a supermarket lorry into the path of a pizza delivery bicycle, which almost runs into an old lady with a shopping trolley. She hears her wolfpack howling in the distance. It's time to return to her family.

Môrwen walks alone through the Wentloog Levels towards Cardiff. Her people left footprints on the banks of the chocolate-coloured Usk, long before the Romans dug *reens*, drainage ditches, to channel the water to the sea. A starling murmuration wheels through the sky where people gathered coal washed down from the mines by the salmon river.

The Salmon Children/Plant yr Eog

Two children lived in a cottage in the middle of Y Coed Duon, the black wood by the Rhymney River. Their father was a woodcutter who taught his children the language of the birds.

One evening they were playing on the banks of the river when an old lady appeared. 'Would you like some fresh gingerbread?'

If only the children had known about weird old ladies in fairy tales, but they followed her to a cottage, where they filled their bellies with gingerbread people. When they had eaten everything, the old lady waved a hazel stick in the air, and chanted, 'Swash, swish, now you're fish.'

Their arms turned to fins and their legs to tails, their bodies grew silver scales with spots and they transformed into salmon.

The old lady placed the salmon children in a basket and poured them into the river. They swam down to the Severn Estuary and into the sea, where they met dolphins, whales, jellyfish and seals. They spent the winter on the seabed alongside weird creatures with rows of sharp teeth and lights on their heads, who taught them the language of fish.

In spring, they swam back to the estuary and up the Rhymney River, where they were caught in a net by the little people who lived behind the waterfall at Pwll y Bel. A girl dressed in green with primroses in her hair and a snake by her side led the salmon children along a silver path of water to their father's cottage. When he saw two plump pink fish, he

rubbed his hungry belly and fetched the frying pan. Primrose Girl transformed her snake into a hazel stick, tapped the salmon on their tails and turned them back into children. And the woodcutter was so relieved he didn't eat his own flesh and blood.

Primrose Girl went to the gingerbread cottage and tapped the old lady on the back of the hands with her hazel stick. Her arms and fingers grew long and twisted till they looked like tree branches and she became a weeping willow, the guardian of the salmon who swim across the Atlantic.

13

Temperance Town
and Tiger Bay

Splott Beach

Môrwen tiptoes through the sea-smoothed bricks, reinforced concrete and rubber tyres – memories of the biscuit factory, steelworks and international airport that once overlooked Splott Beach. She calls into Oasis to see her friend Cath the storyteller, who introduces her to the recent refugees, migrants and asylum seekers who bring inspirational stories from across the world. A Nigerian girl says she wants to be a mermaid when she learns how to grow her tail. Môrwen whispers a secret in her ear about Mami Wata, who swam alongside the slave ships from West Africa. When she comes by next week, the girl will be *Môrforwyn Caerdydd*, the Cardiff mermaid.

Môrwen walks down Viking Road towards the Norwegian Church, built in 1868 for sailors whose ships brought timber from Norway for the Welsh coal industry. Roald Dahl was baptised here by his Norwegian mother Sofie, who told her son the troll stories that inspired the tale of orphan Sophie and the BFG.

The opening of the Suez Canal in 1870 inspired more migrants to come to Tiger Bay from Scandinavia, Africa, Asia, China, South America and the Caribbean to work side by side with Welsh dockworkers and sailors.

Somali Stories of Tiger Bay

A Somali story begins, 'A person who has not travelled does not have eyes.' Cardiff has the largest British-born Somali population in the UK. Their ancestors were seafarers who came to earn money for their families to buy livestock back home. Some married Welsh girls and filled jobs that white workers didn't want, shovelling coal into boilers, loading tramp steamers or enlisting in the Royal Navy. They lived in Somali-run boarding houses and built a strong community that stood up to racism, gang culture, and unemployment caused by the Depression.

Abdi Langara, a Warsangeli, ran a boarding house on Millicent Street for Somali migrants like Ibrahim Ismaa'il, who arrived as a teenager in Cardiff around 1900 to look for work in the docks. Abdi acted as an agent who helped Somali people protect their money so they couldn't be robbed by white gangs frustrated at the lack of jobs for those who survived the First World War. During the anti-immigration race riots in 1919, Ibrahim and other Somalis defended Abdi's house when it was doused with paraffin and set on fire after a fight that left many wounded on both sides. Somalis were put on trial, but freed after the courts ruled they acted in self-defence.

Following the outbreak of civil war in Somalia in the 1980s, families came to join their relatives who were working as solicitors, civil engineers and in the NHS. They built the Al-Noor Mosque on the site of the old Peel Street Mosque, which was demolished during the Butetown development. The old Somali sailors say they stayed because the locals didn't know how to shovel coal properly and someone had to show the kids how to do it.

Betty Campbell

Wales' first Black headteacher, Betty Campbell, was born in Butetown and raised in the poverty of Tiger Bay. When she went to school, she studied alongside mostly white middle-class girls, and grew to love books, especially Enid Blyton's. She became a community leader, race relations campaigner, rule-breaker, bingo caller and teacher in Butetown who wanted her pupils to know their history. She met Nelson Mandela on his only visit to Wales and travelled through North America to learn about Harriet Tubman and civil rights activism, which enabled her to turn her school into a unique centre for multicultural education.

Henry Box Brown

Môrwen leaves the soulless Cardiff Bay barrage, and cuts through the backstreets past Guto the storyteller's mural and Bruno the tobacconist's Bear to the Principality Stadium. In the early 1860s, this was Temperance Town after Colonel Edward Wood sold his swampland to Jacob Scott Matthews, a market gardener who built houses for working-class people, but without pubs.

Time retreats and Môrwen finds herself staring at a poster outside the Temperance Hall advertising 'The King of all Mesmerisers', 'Professor of Animal-magnetism' and 'Doctor of Electro-Biology'. An elegant gentleman appears from nowhere, holds out his hand with a nail in his palm, closes his fingers, and opens them to reveal an acorn, which he tells Môrwen to plant and it will grow into a nail tree.

'You're a magician?'

'Ma'am, I am an African conjurer and stage prestidigitator. Henry Brown at your service. You may have heard my story? I was born into slavery in Virginia and escaped by posting myself in a wooden box to the Philadelphia Vigilante Society. I wrote my autobiography, *The Narrative of Henry Box Brown*, and with

the money I built a moving panorama, 'A Mirror of Slavery', consisting of forty-nine paintings by the abolitionist Josiah Walcot. I took to the stage to tell the stories of my people from their homes in West Africa to lives of slavery in North America, and on to freedom. For the first time, people saw our stories in moving pictures. I am performing my 'crankie' in the miners' institutes and clubs in Aberdare, Merthyr, Monmouth and all round the Valleys, and here I am in Temperance Town.'

'Do you know Mami Wata?'

'Magnificent lady. My heroine. She swam alongside the ships to protect my people. You swim with fishes, too, I see?'

'How d'you know?'

Henry reaches behind her ear and opens his hand to reveal a limpet, some seaweed and a mermaid's purse. He gives her a matchbox painted to look like a tiny toy theatre with a moving scroll that tells the tale of the black mermaid.

As Henry doffs his hat, a dove flies out, and he vanishes, but not before he advises Môrwen to step through the portal on the River Taff.

The Taff Portal

Môrwen passes Clwb Ifor Bach, named after the twelfth-century Welsh hero who protected Senghenydd in upland Glamorgan from Norman invaders, and whose tomb is guarded by stone eagles. Ifor could never have known that Welsh legends like Gwenno and the Super Furries would play in his club.

Môrwen walks along the Animal Wall carved by William Burgess in the late 1800s, and strokes the stone eagle before she follows the meandering river past multicoloured graffiti, tyre dumps and burnt-out cars to the Portal. It's known as one of the Seven Wonders of the World. Well, of Glamorgan.

The Portal is thought to be bottomless and home to supernatural creatures, including a Serpent that drowns people and sucks their bodies dry, a Siren who lures sailors to their deaths, a Snake Girl who enchants lovers, and Bella, who drowned here and haunts the river as a wraith in search of a cwtch.

This is weird Cardiff.

Môrwen dives in, hoping the Portal will take her somewhere lush. The Serpent swims towards her with jaws open, the Siren soothes it with song, Snake Girl wraps herself round its mouth to prevent it biting, while Bella takes Môrwen's hand and leads her out of the Portal. As she breaks surface and gulps the air, she is in the exotic fairy tale world she dreamed of – by the Ferris wheel in Barry Island Pleasure Park.

Selling the Weather

Old folk sell weather in Barry.

Modryb Sina says there will be fair weather between Penarth and Sully, just now in a minute.

Ewythr Dewi sells sunshine as far as Swansea, but only to Welsh speakers.

Bill O'Breaksea offers clear weather to English sailors and traders between Aberthaw, Minehead and Bristol.

Môrwen says she prefers west Welsh drizzle. It's free.

Before the amusement arcades, visitors came to Barry to cure eye ailments by dropping a pin in the healing waters of St Barruc's Well. It must have worked because a local innkeeper kept a pint pot on his bar filled with pins from the well.

One day, St Barruc lost St Cadoc's Bible when his boat overturned in a storm. That evening when Cadoc sat down for dinner and cut open a freshly caught salmon, what do you think popped out? Guts? Nest's wedding ring? No, his Bible, of course.

Dinosaurs

Early in the Mesolithic, the Severn estuary was mostly forest and salt marsh with winding rivers, a great lake and low hills where Môrwen's people foraged and fished. The children played Brontosaurus vs Tyrannosaurus Rex at The Bendricks and stepped in the fossilised dinosaur footprints many times the size of theirs. None of the Mesolithic children had ever seen a dinosaur, but they knew all about them from Nan's stories. In 2005 The Bendricks footprints were cut from the stone, and archived in the National Museum.

Flat Holm Sanitorium

The islands of Flat Holm and Steep Holm appeared when the meltwater from the ice sheets began to fill the Bristol Channel and created a large lake and later an estuary. St Cadoc lived as a hermit on Flat Holm before smugglers moved in and stored tea and brandy in the old cave that led to the sea.

In July 1883, the steamship *Rishanglys* dumped three cholera-ridden sailors on Flat Holm. The islanders isolated them in a canvas tent, and asked Cardiff council for protection, so a sanatorium was built to keep mainland Wales free of plague. It closed in 1935, but the creepy ruins are still there, along with disused gun batteries, a foghorn station, peregrine falcons, black-backed gulls, slow worms and memories of Guglielmo Marconi, who sent the first telegraphy message in Morse code from a transmitter on the beach to another at Lavernock Point between Barry and Penarth.

Swan Girls of Whitmore Bay

Long before Butlin's holiday camp opened on Barry Island, two swans lived at Friar's Point in Whitmore Bay, where they removed their wings and feathers and transformed into girls.

'We loves it when we has a swim,' they said.

One day they were seen by two boys, one from Trwyn y Rhws and the other from Cadoxton, who were shooting wildfowl.

'Wicked, epic, tidy, smart,' said the boys as they stole the girls' wings and feathers.

The girls told them not to be chopsy and give back their clothes.

'We will if you marries us.'

The girls agreed thinking they would soon find their clothes and fly away.

As time passed, they had children with long curved necks who preferred to swim than go to school. One day, the swan-wife of

Cadoxton was hit by a wagon and as she lay injured, the swan-wife of Rhws told her husband to return their wings and feathers so her friend could die as a swan. The two boys finally understood the cruelty in keeping swans as wives, so the girls dressed in their clothes and flew away over the nuclear power plant at Aberthaw. And they're still flying to this day, free as birds.

Jessie the Tattoo Artist

Jessie Knight, the Lady Tattoo Artist of Barry, was born in Croydon in 1904 to a mother she described as a mad alcoholic, and learnt tattooing from her father, who was a sailor, circus performer and sharpshooter. Jessie joined him in the big top as a bareback stunt rider and sharpshooter's dummy, but after he accidentally shot her twice when she was 18, she gave up the greasepaint and became a tattoo artist. She married at 27, but her abusive husband disliked her art, and when he kicked her dog down the stairs, she used her sharpshooting skills on him, and ended the marriage.

Jessie worked in her father's studio in Y Bari, sketching freehand with coloured pencils and a matchstick dipped in ink to draw on to the client's skin. Her designs of sailors and near-naked women were so popular they were often stolen by rivals, so she kept her artwork hidden in a trunk, which she sat on while she worked. She was the only professional female tattoo artist in the UK, and turned down many proposals from her customers because she made this previously male art form accessible for women, who felt more comfortable in her company.

Ceffyl Dŵr of Aberthaw

One dark night with snow in the air, long before the decommissioned coal-fired power station was dreamed of, an old man walked over the moors towards Aberthaw. As he crossed the beach, he saw a long-legged man sat on a glowing white horse, illuminated by the full moon. He followed the horse and rider up to the ruins of the old lime kiln, where they vanished into the darkness. He curled up and went to sleep, and when he awoke he saw the whole valley had been flooded by the tide. Had the water horse not guided him through the darkness to safety, he would have drowned.

Môrwen climbs on the back of the water horse and they leap over the crumbling cliffs of Porthkerry, and fly west along the coast towards St Donats.

And she has a lush tattoo of a mermaid on her arm.

14

The Lady of Ogmore and the King of Porthcawl

LLANTWIT MAJOR–ABEROGWR–PORT TALBOT

The Sad Witch

As Môrwen is carried over Llantwit Major by the Water Horse, she hears the saddest sound – an eerie wailing followed by a long sigh, then silence. The village opens its windows to see a woman standing motionless in St Illtyd's Churchyard, dressed in a frayed black gown, with a ghostly face, piercing grey eyes, yellow teeth, and wild red hair. She flaps her leathery arms and drifts out to sea weeping, '*Fy ngwr*, my husband!' In the morning, a ship lies wrecked on the shore, the crew lost to the sea.

She is the Sad Witch, *Gwrach y Rhibyn*, heard only before a deadly storm. Emma Thomas, who wrote down the story, was born in the village in 1853, daughter of stonemason and antiquarian Illtyd Thomas. She married a French journalist, Louis Paslieu, but left him when she discovered he was already married, and raised her daughter Bronwen as a single mother in Llantwit Major. She worked as a fortune teller, Madame Paslieu, and wrote books about stirring stories of the sea, rip-roaring Arthurian legends and a gentle Quaker romance. In 1909 she published a phantasmagoria of folk tales under the pseudonym Marie Trevelyan, based on her father's collection of Glamorgan folklore mixed with his daughter's imagination. Her approach was very different to the textual analysis and comparative mythology beloved by male antiquarians and clergymen.

Water horse flies westwards over St Donat's, where there was once a storytelling festival with giant puppets, and into the darklands of Dunraven, where human remains were found beneath the beach in 2024.

The Dunraven Wreckers

Walter Vaughan, Lord of Dunraven, had watched many ships wrecked off the coast at Trwyn y Witch, so he wrote to the government offering thoughts on improving safety at sea, but his ideas were continually rejected. His wife died of a broken heart, two sons were lost in a storm when fishing, another drowned in a vat of whey, while his eldest son left for a new life abroad but his ship went missing.

Walter turned to the dark side. He plundered a beached cargo ship off Dunraven to pay his debts, and joined forces with the leader of the local wreckers, Mat of the Iron Hand. They met when Mat was on trial before the local magistrate, who happened to be Walter Vaughan, and they recognised each other's tormented spirits.

One night during a storm, Walter and Mat watched as a man swam ashore after his ship tore itself apart on the rocks. No survivor was allowed to live, so Mat snapped the man's neck and cut off his ring hand. When he showed it to Walter, the Lord of Dunraven's torment was complete. The rings on the fingers belonged to his eldest son, who had left home years before to escape his bitter father and make his fortune.

The Lady of Ogmore

The Norman Lord of Ogmore had caught a Welshman poaching on his estate and was about to imprison him, when his daughter spoke up:

'Father, it's my birthday. May I have my present now?'

'Of course, my dear. What would you like?'

'This man's freedom.'

'I'm sorry, but he was caught poaching on my land.'

'This is not your land, father. We took it from this man's ancestors. And one day the sea will take it from you.'

'Nonsense, that is dangerous thinking. I own this land!'

'Land cannot be owned, bought or sold. '

'I didn't buy it. It was given by my father.'

'And where did he get it from?'

'From his father.'

'And your grandfather?'

'He won it in battle.'

So the Lady of Ogmore spat on her hands and rolled up her sleeves. 'Right. I'll fight you for it, father!'

The lord laughed at his daughter's impertinence, but said she could have all the land that she could walk around barefoot before sunset. She stared into her father's eyes, kicked off her sandals, and began to walk. The thorns and flints tore her feet, but she followed the coast and rivers, with her father's soldiers behind her armed with tape measures, until the sun set in the sky. The lord kept his word, and the land his daughter walked around was returned to Wales as the Merthyr Mawr Nature Reserve.

The Shee Well that Ran Away

Water Horse drops Môrwen by the ruins of Ogmore castle, and vanishes into the river where the Ewenni and Aberogwr meet. Môrwen spits three times and crosses by the stepping stones used by the Lady of Ogmore to visit her lover in Merthyr Mawr.

Girls washed their underwear in the Ewenni river where three springs met at the Shee Well. The girls knew if they carried their underwear home between their teeth and placed it by the fire to dry, they would see their future partners in the flames. And if it wasn't the person they hoped for, they simply turned their underwear round to make them disappear.

The girls knew lots of stories and sayings.

'Never wash new clothes on the new moon or they won't fit properly.'

'Clothes washed right-side-out will quickly lose their colour.'

'Pull your underwear from the tub upside down and you'll never be witched.'

A Water Ogre lived near the Shee Well, where it ate all the fish, left its mess in the river, and stole the girls' underwear. The Shee Well was fed up with the Water Ogre, so it ran away and hid in a cave in the hills, taking the river water and all the fish with it. There was nothing left for the Water Ogre to eat or drink, so it pleaded with the Shee Well to return. The Well agreed on condition the Water Ogre cleaned the river, looked after the fish and stopped stealing the girls' clothes.

So the Shee Well returned and if anyone makes a mess in the river, they will have to answer to the Water Ogre.

Elfys, King of Porthcawl

Môrwen follows her people's footsteps along the eroding cliffs at Nash Point, past the coloured caravans of Trecco Bay to Porthcawl's Coney Beach funfair, named after American soldiers who were stationed there. She meets a dark-haired man with a quiff and a crooked smile, dressed in a denim jacket, drainpipe trousers and blue suede shoes.

'Excuse me, ma'am, but would you like to be serenaded?' he drawls.

Môrwen rubs her eyes. The beach is full of Elvises. Porthcawl has hosted the Elvis Festival since 2004 and there are now more than 800 of them, many with Valleys accents, who compete for the best

Gospel, best Vegas and best GI Joe. One was Peter Singh, who owned a takeaway in Swansea that sold 'Love me Tender Burgers' and meals for one called 'Are You Lonesome Tonight?'

'Are you Welsh?' she asks her Elvis.

'Child of the Preseli Hills, descendant of St Elwys of Haverfordwest, son of Gwladys and Vernon, and soul brother of Mr Tom Jones of Las Vegas.'

She stares at him. 'Are you a real king?'

'I ain't nothin' but a hound dog, ma'am. The real King runs Ernie's fish and chip shop on Portland Road in Aberystwyth. Everyone knows his secret, but they don't tell because he batters the finest fish in Wales. He signed his name on a rock at Eisteddfa Gurig.'

'I saw him busking on the prom when I was in Aber!'

'He will always be on our minds,' mutters Porthcawl Elvis as he walks away strumming an out-of-tune chord on his guitar:

Well, since my baby left me
I found a new place to dwell
Down at the end of the Esplanade
At the Grand Pavilion Hotel.

'Viva Porthcawl!'

The Devil's Ship

The Devil kept the souls of sinners in a three-masted ship anchored off Porthcawl. It smelt so strongly of sulphur that it offended the nose of a giant, who used the mast for a toothpick, wore the wheel as an earring, blew his nose on the mainsail, pierced the hull with a spear and sank it.

There are more giants in Porthcawl. On 24th August 2004, the Lifeboat saved two fishermen from 10ft waves called the 'merry dancers' off Nash Sands. Crew member Aileen Jones became the first woman to be awarded an RNLI medal in 116 years.

The Submerged Castle of Cynffig

Another Norman lord like the one at Ogmore built himself a castle on Welsh land at Cynffig and introduced laws to keep the unruly people under control. He prohibited fighting, wrestling and gambling, taxed beer and bread 'til it was too expensive, and if a woman was found guilty of witchery by a jury of six men she would be tied to the ducking stool for an hour. So a Welsh chieftain cursed the Norman lord that after nine generations the land would be returned to Wales.

In the blink of a crow's eye, nine generations passed, and at midnight a voice was heard in the castle: '*Dial a ddaw!* Vengeance is coming!' The fire flickered in the hearth, lights went out, a sandstorm covered the land, the rain fell and the castle was flooded beneath Kenfig Pool. The land is now Kenfig Dunes National Nature Reserve, with its rare fen orchids, booming bittern and Mesolithic footprints in the peat beneath the sandy beach.

The Maid of Sker

Elizabeth, daughter of Isaac Williams of Sker, loved nothing more than dancing all night at the Prince of Wales Inn while the harpers played and the old women gossiped about the mischief they got up to when they were young.

One Gŵyl Mabsant, Elizabeth was dancing wildly to the playing of a young harper from Newton-Nottage called Thomas Evans. She couldn't take her eyes off the way he moved to his music, and by dawn they were lovers.

In the morning Thomas wrote a poem for her, *Y Ferch o'r Scer*, The Maid of Sker, and his friend David Llewellyn of Nottage set the words to the hymn tune *Diniweidrwydd*.

When old Isaac Williams heard about his daughter's lover he was furious. He was a gentleman farmer with a reputation, and he didn't want an itinerant musician for a son-in-law, so he set the dogs on Thomas and informed Elizabeth she was to marry a Mr Kirkhouse of Neath.

The marriage of Mr and Mrs Kirkhouse took place at Morriston, and they settled at Briton Ferry, where Elizabeth bore four children. All the while she continued to see Thomas the harper in secret, until after nine years of being torn between her children and her true love, Elizabeth's heart broke. She died aged 34 and was buried in Llansamlet.

In 1872, R.D. Blackmore, author of *Lorna Doone*, wrote a bestseller, *The Maid of Sker*, a tale of mysterious orphans and improbable shipwrecks that bore little resemblance to either truth or folk tale.

Thomas waited till he was 50 before he married and had children. He lived to be a grand old man, until one day in 1819 he collapsed while playing bando at Newton.

Margam Bando Boys

Spanish coins from a shipwreck have been found on Margam Sands, the beach where the Bando Boys ruled. Bando was like hockey, played with curved sticks of willow and ash with the intention of knocking a leather ball through a goal or into the neighbouring church porch. Rules were few, bets were taken, and the ball forgotten as blood flowed and bones broke while thousands watched. Once a girl hid the ball beneath her skirt, ran to the player she fancied and dropped it at his feet. Llantwit Major were feared, Pyle fancied themselves, while Margam sang about how they defeated Napoleon and took Paris for Wales, and this was long before rugby was invented.

The Fairy Tale of Port Talbot

Môrwen walks past the silhouettes of the blast furnaces and twisted chimneys that symbolise Port Talbot's sense of strength, resilience and radicalism. It has already witnessed the second coming thanks to the man she passes on Aberavon Beach, who greets her with a booming voice and a sheen of familiarity. Another legendary hero lives here, Captain Beany, founder of the Baked Bean Museum of Excellence, which was housed in the living room and kitchen of his flat in Sandfields. It closed recently when the Captain retired, though he still has sixty baked beans tattooed on his head.

She drifts across the River Neath beneath the M25 at Briton Ferry, where smuggler Catherine Lloyd was landlady of the Ferry Inn. Ponies graze on a travellers' site unseen by motorists, as are the fen raft spiders and shrill carder bees of Crymlyn Bog. When she reaches the outfall pipes and urban dumps beneath Swansea Bay, Môrwen hurries towards the city to meet the woman who first greeted her when she stepped through the veil. Myra the storycatcher.

15

Swansea Jack and Potato Jones

Myra Evans

The ghosts of the Swansea Cape Horners sing sea shanties and squeeze concertinas as Môrwen walks through the old ferry port to the Maritime Museum, where Myra is waiting. She is no longer Myra Rees, having married her childhood sweetheart Evan Jenkin Evans, a nuclear physicist from Llanelli. She is Myra Evans, teacher, writer, illustrator, folklorist, storyteller and mother of six.

In the early 1920s Evan found work as a lecturer in Physics at Swansea University, so the family moved to suburban Sketty, with its semi-detached villas and stained glass windows. Myra worked in primary schools in Barry and Swansea, teaching Welsh through folk tales like Gelert, Glyndŵr and Gwenllian, which she wrote out in exercise books for her students. She compiled and illustrated the first Welsh language primer, which was published by Foyles of London, and she told nature stories in Welsh every Friday on BBC Radio in Cardiff as 'Anti Myra'.

Aneurin, her eldest boy, went to Swansea Grammar School with his bestie Dylan, the mischievous son of Myra's neighbours in Sketty, David and Florence Thomas. Dylan had a head full of golden curls, eyes of autumn brown, and the disadvantages of a happy childhood. He was molly-coddled by his mam, stole sweets, fell out of trees and was as cheeky as any changeling child.

He wanted to be a bohemian poet, looked after by soft silly ravens. He had the qualifications. He was an outsider who wrote about himself, rather than the romance of Welsh mythology.

By 1934, Dylan had lost his jobs at the local newspaper and the Swansea Little Theatre, his first book of poetry had been accepted by a posh London publisher, and he had a girlfriend in Battersea. He left the comforting arms of the Mumbles Mermaid and crossed the border into the neighbouring kingdom, where he slayed the giants of the literary world and charmed the dragon, Dame Edith Sitwell, who reviewed his book favourably. He attended the court of King John, the painter Augustus, where he wooed the golden-haired Princess Caitlin with words, not swords. They married and had children, and his friend Geoffrey Grigson called him 'the Swansea Changeling', left by the Otherworld under a foxglove.

Swansea Jack

A hairy black retriever bounds up to Môrwen, 'Is this your dog, Mrs Evans?'

'He's William Thomas' dog. Swansea Jack,' says Myra. 'They live in Roger Street, Treboeth. In 1931 they were walking through the Docks when Jack saved a 12-year-old boy from drowning, though no one knew about it as the boy was too scared to tell his mam. A few weeks later he rescued a swimmer and this time it was all over the papers. Jack rescued twenty-five more people and travelled the country raising money. He became so famous that the *Star* voted him the Bravest Dog of the Year in 1936, and the council awarded Jack a silver collar.'

Myra and Môrwen link arms and walk down Castle Street telling tales while they flip through each other's sketchbooks. In the Swansea Valley, myths and fairy tales are woven into conversation. '*Chwedl Cymraeg?*' means, 'Do you speak Welsh?' and also, 'Do you tell stories in Welsh?'

Jack barks in Swansea Welsh.

The Kardomah Cafe

Jack leads Myra and Môrwen into the Kardomah Cafe. There are rows of Formica tables, a chequerboard floor, sci-fi coat racks, mosaic-tiled columns and dark wood-panelled walls. A well-posh woman sips from a Kardomah cup with her pinky out. A man in a flat cap plays '*Carreg Bica*' on his flute. And a curly-haired young man is writing in a notebook. He waves to Myra.

'Welcome to my My Home Sweet Homah, Mrs Evans. Pull up a chair.' Dylan sees Môrwen. 'And you're Mesolithic Girl? You're 7,000 years old? Tell me your story. I'll buy breakfast? I have a tab here.'

Myra laughs and rolls her eyes.

A waitress says 'What d'you fancy today, lovelies? We have eggs and bangers from the Gower butcher, fresh salmon from Coakley Greens, and fried tomatoes from Swansea market.'

'I'll have it all,' says Dylan. 'Mrs Evans will pay.'

'She won't!' Myra winks at the waitress.

'I'll add it to your tab, lovely boy,' the waitress grins.

She brings Jack a bowl of water, coffee for Myra, hash browns for Môrwen, and a full Welsh breakfast for Dylan, who spills egg down his shirt. A cold air fills the room. Standing motionless around the table are wraiths. William the Scabby, rebel Welsh warrior born in 1262, captured by the Normans and hung twice on a gibbet on a nearby hill, but miraculously never died. Ann of Swansea, born into a poor theatrical family, became a model in a brothel where she was shot in the face, found success on Broadway and after several scandals. She settled in Swansea, ran a bathing house and wrote fourteen Gothic novels, including *Cambrian Pictures or Every One Has Errors* in 1810 about a cross-dressing Welsh girl called Adeline. Swansea is a city of literary ghosts, where true and tall tales flow like Kardomah coffee.

Snake Girl

A Swansea girl with eyes that shone like emeralds had mesmerised a young farmer from Ynys Môn, who invited her to live with him. She agreed, providing she was free to leave the cold north twice a year. He assumed she wanted to visit her family in Copper Town, so he agreed.

All seemed well, until his mother became curious, as mothers often do. She suspected the girl had a lover in Swansea, so she told her son to follow. So when the girl left, the farmer trailed her through the woods and along the river to a deep dark pool. He hid behind an elder tree and watched as she removed the green girdle from her waist, threw it onto the grass, raised her arms in the air and slowly vanished. He ran to the pool, but all he saw was an emerald snake that hissed at him and slithered into a hole in the bank. He poked at it with a stick, but the snake never emerged.

Two weeks later, when Swansea girl returned, he asked her about the snake. She stared with emerald eyes, licked her lips with a forked tongue, and hissed. For the first time, he had suspicions.

Six months later, when she was due to leave for Swansea again, he hid her girdle. She told him to return it, but he refused and mansplained it was for her own good. Her neck arched, her tongue flickered and she spat venom. Thinking the girdle must be responsible, he threw it into the fire. She screamed and writhed, a fever consumed her, and she vaporised like a will-o'-the-wisp.

Gotham City

In 1954, the Mettoy factory opened in Swansea, where they made die-cast toy cars named Corgi after the small yappy Welsh dog. They produced James Bond's Aston Martin with an ejector seat, and the flying car from *Chitty Chitty Bang Bang*, but the most popular was the jet-black Batmobile based on the fab TV series, with rocket launchers and tiny Batman and Robin figures in the front seats. In 1966, Mettoy sold over 5 million Batmobiles as Batmania spread worldwide. So you see, Swansea is really Gotham City.

Joe's Ice Creams

'Who do you prefer, Catwoman or Captain Cat?' asks Dylan as the literary spirits leave the Kardomah.

The waitress laughs. 'Wear a mask and cape next time you launch a book, lovely!'

The Kardomah had been open for thirty-five years when it was destroyed by a bomb in 1941. It reopened on the corner of Park Street and Portland Street sixteen years later and has been run by the Luporini family since 1970, still with its 1950s retro decor.

As the literary spirits wander past the Swansea Grand Theatre, Myra offers to buy everyone a Joe's Ice Cream, which has been made to Luigi Cascarini's unique vanilla recipe since 1922. As they walk to the beach with ice cream round their mouths, Môrwen thinks Swansea cuisine is way better than foraging for nuts and getting scratched by brambles.

'Catwoman!' she says.

Potato Jones

At the height of the Spanish Civil War in 1937, Davey Jones, captain of the *Marie Llewellyn* in partnership with local landlady Edith Scott, set sail for Bilbao with three tramp steamers loaded with a thousand tons of potatoes to feed the hungry Basques, who were being starved by a blockade of General Franco's ships. The newspapers nicknamed him 'Potato' Jones.

Potato was born in Swansea in 1871, and as a boy he cycled to Mumbles Head to watch ships from all over the world sail in and out of the docks. His party trick was to dive under a moored ship, and while the crew looked over the side for him, he swam beneath the ship, climbed up the other side and scrambled like a monkey up the rigging to the top of the mast, before passing a hat round for half-pennies. By the time he was 15, Potato was a Swansea Cape Horner.

He was 67 when he set sail for Bilbao, alongside two more ships captained by 'Ham and Egg' Jones, and 'Corn Cob' Jones. The Royal Navy refused to protect them, suspecting they were smuggling guns and ammunition to the Republicans, which they were. A Spanish battleship was ordered to sink them on sight, and the Joneses found themselves without communications in a storm that lasted four nights. Newspaper headlines asked, 'Where is Potato Jones?'. Some thought he had joined the 174 Welshmen killed in the Civil War. In fact, the weather had forced him to return to Alicante and dump his cargo. Still, Potato Jones is another Swansea hero.

The Mumbles Sisters

The Prussian barque *Admiral Prinz Adelbert* ran aground here in a fierce storm in the winter of 1883. A steam tug, *Flying Scud*, was sent to pull her to safety but the hawser broke and she hit the rocks west of Mumbles Head. The lifeboat was launched, manned by Coxswain Jenkin Jenkins and four of his sons. They cast anchor next to the ship and threw grappling irons on board and rescued two of the crew. However, as they hauled a third aboard, the sea smashed the lifeboat against the barque, and threw everyone into the water. Coxswain Jenkins swam ashore with one of his sons as the lifeboat capsized, and sought help from Abraham Ace, the lighthouse keeper, and his two daughters, Jessie and Margaret. The sisters had already waded into the sea up to their waists with shawls knotted together to make a rope to haul the injured and drowning men to safety. In the morning, at low tide, the remaining crew escaped, but four lifeboatmen died, including two of Jenkins' sons. Had the lifeboat not launched, they may not have lost their lives, but the RNLI always answers the call.

Mumbles Railway

Môrwen has wiped the ice cream from her face and is bouncing with excitement at the thought of riding the Mumbles tram. Myra holds her hand to prevent her being run over as she climbs to the top deck, spreads out on the front seat, presses her nose to the glass and draws faces in her breath. Dylan sits behind and scribbles ideas on the back of an envelope, while Myra draws the tram driver in her sketchbook. This was the first passenger railway in the world when it opened in 1807, pulled by horses and later by a Puffing Billy.

The Mumbles Railway carries childhood memories of trains running over lost sheep, a drunken policeman killed by an oncoming horse, a local man who lost an arm when the locomotive drove over it, drunkenness on the last train from Southend and passengers on the open decks covered in smut. The driver for the final journey in 1960 was Frank Duncan, who had worked on the railway for fifty-seven years. He once stopped the train to pick up a chicken on the line, and ever since it travelled on the footplate with him.

Myra gets off at Sketty and says she will see everyone in Cei Newydd and treat them to Conti's ice cream. She giggles. *Conti* is a rude word in Welsh.

Dylan jumps off, swearing he saw a vampire clawing at the window. He tells Môrwen to call at his writing shed when she's in Laugharne and he'll treat her to a brown ale at Brown's Hotel.

Jack wags his tail and bounces up to his old friend, Carl the storyteller, who has told the dog's tale all over the city.

Môrwen is alone in Southend. The crowds have faded away. Only a round plaque remembers the extraordinary ordinary people who travelled on the clanky sooty railway.

16

Swan Girl and Water Horse of Gŵyr

GŴYR/GOWER

Curse of the Verry Volk

A Norman lord was celebrating his daughter's wedding at Pennard Castle, overlooking Three Cliffs Bay on Gŵyr, when a guard reported seeing strange lights in the woods. The lord, fearing the Welsh were attacking, gathered his soldiers and went to investigate. In a clearing were the Verry Volk, dancing in celebration of the marriage. The lord, drunk on ale and fear, thought they were mocking him so he ordered his soldiers to attack.

Standing amidst the carved bodies of her dancers, the Verry Queen pointed at the lord, cursed him for his cruelty and cowardice, wailed into the wind and vanished. The sea stirred, the wind blew, and the beach poured over the land like water, until the castle and all its people were drowned in sand.

Construction on Pennard Castle began in the twelfth century, but a survey in 1650 described it as 'wholly besanded' and 'altogether unprofitable'. If anyone spends a night there, they may hear the wailing of the Verry Queen, and if they are from one of Gower's wealthy old families, they will wake in the morning either mad or a poet. Or a golfer. Strange how golf courses are built on top of the Otherworld.

Gower Pony and Water Horse

As the sand blows across Oxwich Bay, a *Ceffyl Dŵr* leaps from the sea on the seventh wave, gallops across the beach and into the churchyard, prances on its hind legs and vanishes through a closed gate into eerie silence.

Môrwen meets a Gower pony grazing on the clifftop and strokes her nose. These ponies wander the common lands in all weather and are tough enough to stand on the marsh until the high tidewater reaches their necks, where they stay until the tide goes out.

Pony allows Môrwen to climb on her back and says, 'My sisters were the Queens of Gower. Listen, and I'll tell you some horsey stories.'

'You can speak?'

'Yes, though I'm a little hoarse.'

'And you tell stories?'

'Of course. All ponies have tails.'

She trots west towards Rhossili.

The Undertaker's Horse

John Beynon is the ninth generation of a family who have lived at Kimley Moor Farm, Rhossili, since 1717. In the 1970s, inspired by the yarns of Cyril Gwynn, Bard of Gower, John began writing folk poetry. 'The Undertakers Horse' was told to him by Ernest Richards of Bank Farm Caravan Site, about a true lifeboat accident in 1912:

Back in the days before tractors, the horse was Queen of Gower. One old Clydesdale called Blossom lived near Rhossili, where she pulled the lifeboat down to the sea whenever she heard the boom or saw the flares. She galloped to the quay so quickly, she helped save many a life.

As Blossom grew older, she slowed down and the lifeboat crew feared she was ready for the Knacker's Yard, but they couldn't bear to think of her ending her days as glue. So they gave her to the undertaker, for she walked at the perfect pace to pull a hearse.

One day, the bearers in their black top hats were about to lift the coffin off the cart at the door of Port Einon Church when the boom went off. Blossom's ears pricked up. She saw the flares. Someone needed help. She trotted as fast as her old legs could carry her, down to the quay to launch the lifeboat, with the coffin bouncing up and down on the cart behind. The funeral continued without the body, though the relatives complained they hadn't ordered a burial at sea.

Old Horse

John Beynon's grandfather kept a horse's head buried in a field at Kimley Moor to be dug up at New Year, and decorated with ribbons, a wire snapping the jaw, and a sheet over grandfather's head. It went round all the pubs on Gower, frightening the hell out of everyone and singing:

Once I was a young horse and in my stable gay,
I had the best of everything of barley oats and hay,
but now I'm getting an old horse, my courage is getting small,
I'm obliged to eat the sour grass that grows beneath the wall.

One year, grandfather got so drunk he couldn't remember where he buried the head, and that was the end of the tradition.

The Red Lady who Became a Man

Horse bones were discovered in Goat's Hole Cave in the early 1900s, alongside a headless human skeleton stained with red ochre, and the remains of bear, fox, hyena, bison, wolf, reindeer, woolly rhino, hairy mammoth and more than 4,000 flints, teeth and seashell necklaces. A male archaeologist thought the skeleton was a Mesolithic witch or prostitute, as she was buried with a sheep bone he believed was a conjuring tool. However, the Red Lady turned out to be a 33,000-year-old young man, from a time when the land was covered by a glacier and the cave was far from the sea.

'Hey, Pony, maybe the Red Lady had two spirits?' Môrwen muses. 'My Nan did.'

'Hay? Where?' asks Pony.

'Hey, not Hay. Are you hungry?'

Môrwen feeds Pony some dried grass from the edge of a field. Pony tells another story:

Swan Girl of Gŵyr

A swan lands on a rock on the coast of Gower, removes her wings and feathers, out steps a girl, washes herself and gives thanks to the sea. All is well in her world, until a young farmer sees her and becomes enchanted. He picks up her wings and feathers and holds them to his cheek. She stares into his eyes and tells him to return her clothes. He refuses, and says he loves her. She tells him he knows nothing about her, they have never spoken, and she isn't from his world. It's infatuation. He locks her wings and feathers in an oak chest at his house by the sea and gives her pretty dresses, which scratch her skin. He feeds her fine foods, but she craves seaweed from the rockpools. He lights a fire to keep her dry, while she dreams of drowning. Every night he unlocks the chest and holds her feathers against his cheek, and every night she watches her sisters fly past the moon. Years pass, until one night he forgets to lock the chest.

She dresses in her clothes, flaps her wings, her arms ache but she throws herself from the window and flies away, and she's still flying, free as a bird.

The Wobbly Ox

At Eynonsford Farm near the sea at Port Einon, stood a thatched longhouse with a kitchen at one end and a cowshed at the other, where the farmer kept his prize ox. One summer night he heard music coming from the cowshed, so he shone his lantern inside and there were the Verry Volk, dancing on the back of his prize ox. When the music stopped, the ox dropped down dead. They swarmed over the body and skilleted it into a thousand steaks, leaving the bones on the floor. They lit a fire outside, roasted the ox and feasted through the night. Then they reassembled the bones into the shape of an ox, stretched the hide over them – and there was the farmer's ox, alive as ever. Except one of the bones from its foreleg was missing. They searched until the sun rose over Cefn Bryn, and then vanished. When the farmer returned to the cowshed in the daylight, he found his ox with one foreleg longer than the other, walking with a lopsided limp.

Stints

Pony carries Môrwen along the beach and cliffs, past the 18m-high stone wall with lopsided windows at Culver Hole, built in the thirteenth century as a pigeon coop and later used by smugglers from the Salt House at Port Eynon. Further along the coast at Rhossili, the ribs of a sea serpent protrude through the sand, the remains of the *Helvetia* wrecked while transporting timber from New Brunswick to Swansea in 1887. At Burry Holms is a tidal island, once a hill on the plain now submerged beneath the Severn estuary. Mesolithic people lived here and made birch bark tar, a sticky resin used as a glue.

On Cefn Bryn, each farmer has a 'stint' that allows them to graze their animals on the common land in proportion to the acreage of their farm. So one farmer can graze a dozen sheep and two cattle on Llanmadoc Hill, while another can keep a couple of ponies on Cefn Bryn.

Willie John's Fairy Beer

Willie John lived with his mam in a cottage at Llanmadoc, where he brewed sea-salty beer that only he could drink, because it tasted of seaweed. One evening Willie's mam was feeding the hens when she saw the Verry Volk standing amongst the primroses. They explained that they wished to reward the villagers for their kindnesses by giving them a few grains of gold dust, but they needed to borrow a sieve to separate the grains from the nuggets. They pointed to the sieve that Willie used for straining his beer. Mam lent them the sieve and off they went over Llanmadoc Hill, without a thank you. 'You're very welcome,' she said. Later that night there was a tapping at the door, and there was Willie's sieve leaning against the wall.

Later that summer, Willie settled down in his chair by the fire and took a sip from a newly brewed cask of beer. Soon there was a twinkle in his toes, and he danced around the room to imaginary fiddles till his trousers fell round his ankles and his mother had to carry him up to bed. Every night this happened, and after only a very small tankard of beer. This was the strongest and finest ale he had ever brewed, and what's more, the cask never emptied. It was a come-and-come-again beer.

Soon Willie John's beer was the talk of North Gower, and the pubs of Cheriton and Llangennith emptied as the old farmers and rowdy cockle girls danced all night to Phil Tanner's mouth music, and Willie and his mam had more money than they ever dreamed of. Then mam got to thinking. She realised the Verry Volk had enchanted the old sieve before they returned it.

Of course, knowing the Verry Volk's business is to break the spell, and the day came when there was no more beer. Until Willie's mam was cleaning the cobwebs from the brewery and found a little bag of gold dust, just enough to buy hops for Willie to continue brewing the finest beer in Gower.

Penclawdd Cockle Women

One day, back in the early 1800s, the cockle women from Penclawdd, dressed in their aprons, shawls and wide-brimmed hats, were hurrying home from Whiteford Sands with donkey carts laden with cockles. The incoming tide was moving faster than their legs, when they heard the sound of galloping hooves and saw an enormous woolly rhino creature with a large horn on its nose, charging towards them from Broughton Bay. It upturned the carts, attacked the donkeys and chased the women back to Penclawdd. Over the following months the creature reappeared on Llanrhidian Sands, and soon no one dared to gather cockles on the exposed sands.

The women's families relied on the income from cockle gathering. Some of their men were out of work, injured or disabled, so they boiled the cockles in their back garden sheds after opening the shells with scrapers made from old reaping hooks. Then they carried the cockles in baskets balanced on their heads to sell at Swansea Market.

So they asked an old *dynes hysbys* from Cheriton about the woolly creature. That evening the old woman walked through Whiteford Burrows to the beach, drew a large circle in the sand with a ram's horn, and made a geometric pattern with dead-man's fingers and laverbread. The moon shone brightly as she muttered an incantation, summoning the creature, who scratched at the sand and charged into the circle. It stopped in front of the old woman, snorting and stamping. She ordered it to go and never return to Llanrhidian Marsh or Whiteford Sands until a hundred thousand tides had ebbed and flowed. The creature calmed and the sound of its hooves melted into the mist as it vanished forever.

A hundred thousand tides have ebbed and flowed across the marsh since then, so the Woolly Rhino of Whiteford may soon reappear to chase the 4x4s that the cockle gathers now use, having pensioned off the donkeys.

Pony walks slowly into the fast-rising water near the cast-iron Whiteford lighthouse, and stands still as the tide races in around her. Soon she is asleep with Môrwen on her back and only her head above the high tidewater.

17

Amelia and the Hell Dog of Laugharne

LLWCHWR/LOUGHOR–LLANSTEFFAN–
TALACHARN/LAUGHARNE

Amelia

Môrwen slides off Pony's back and swims across the Loughor Estuary towards the coal town of Burry Port, when a small orange Fokker F.VII seaplane named *Friendship* flies over her head and lands on the water, scattering the wintering wildfowl. The cockpit door opens, a woman throws out a rope and motions to Môrwen to tie it to a floating buoy.

'Thanks, sweetie. Where are we? Southampton?' asks the woman.

'Pwll. Wales. *Croeso i Gymru. Môrwen i fi.*'

'Thanks honey. Sorry, I don't speak your language. I'm Amelia from Kansas. We've flown from Newfoundland and these guys are Bill and Slim.' The two pilots wave cheerily.

Môrwen has never seen a plane before, but she wants one.

It's 18th June 1928 and Amelia Earhart has become the first woman to fly across the Atlantic, though Bill did the driving.

A man in a coracle rows out to help, and the plane is towed to Porth Tywyn, where a huge crowd has gathered. Amelia and her companions spend the night in the town while *Friendship* is refuelled and they continue to Southampton in the morning.

Almost a century passes. Môrwen is standing outside the Neptune Hotel opposite Pembrey Burry Port Railway Station, thinking about catching the train along the estuary to Ferryport, when she hears a voice behind her, 'Honey? Thanks for your help back in 1928. I guess it's still Halloween? Spirits and mischief, huh? Can I buy you coffee? There's a diner named after me, and I wanna go.'

Amelia hugs Môrwen and they walk through Penbre dunes between the railway line and the estuary, leapfrogging the groynes and climbing the skeleton ribs of *La Jeune Emma* that

poke through the mudflats like the Rhossili sea serpents. *Emma* was wrecked here in 1828, after being lured to lights placed on the beach by *Gwyr y Bwyelli Bach*, the Little Axe Men, the Cefn Sidan wreckers. Adeline Coquelin, the 12-year-old niece of Napoleon Bonaparte, was on board, and is buried in St Illtyd's Churchyard in Penbre.

Amelia relaxes with a cappuccino at Amelia's Vintage Tearoom near Penbre airport. 'Y'know, when I landed here in 1928, women over 21 had only just got the right to vote? Yet I flew solo across the Atlantic just four years later. Crazy, huh?'

Môrwen nods, 'I was born 7,000 years ago, and men think they're boss. Yet women can fly like birds.'

'I love you, babe. Tell me a story?'

'Sure will, ma'am.' Môrwen is learning Amelia's language.

The Water Horse of Tywi

On the banks of the Tywi lived a dapple-grey *Ceffyl Dŵr* with eyes of fire, a snort like dynamite and hooves that faced backwards. If anyone tried to ride her, she jumped into the water and drowned them. A farmer managed to bridle her and fasten her to his cart, so she dragged them into the water.

One day she rescued a coracle fisherman and gave him a ride home on her back. She loved coracles.

Spookiness Near Llanelli

Two miners from Cydweli, one scarred and one lame, were dangling their feet in the waters of Pistyll Teilo, when the stream spoke. 'It is cold and lonely waiting, waiting, waiting for the sons of William, William, William ...' They hadn't a clue what this meant, but one thought it was a *bwca* while the other said it was *ysbrydion*. Either way it was scary and ghostly so they ran. When they reached home, they realised the lame man could run and the scarred man had skin soft as a baby's.

The Headless Princess

Gwenllian was born in 1097 in Aberffraw, Ynys Môn, daughter of the King of Gwynedd. She married Gruffydd ap Rhys, King of Deheubarth, and they lived at Dinefwr, where they led the rebellion against the Norman invaders. When Gruffydd travelled to Ynys Môn to invite Gwenllian's father to join the rebels, the Normans attacked. Gwenllian raised a Welsh army and marched on the Norman stronghold of Castell Cydweli, but in the battle both of her sons were killed and she was beheaded.

As time passed, a ghostly headless woman was seen on the slopes of Castell Cydweli, searching for her head along the estuary. After three days without luck, she picked up a large stone, held it under her arm and placed it on her grave. The hillside was never haunted again, until the mid-1970s when King Arthur appeared at the gates of Castell Cydweli and oppressed a peasant, but that was in *Monty Python and the Holy Grail*.

Friendship

'What happened to you?' Môrwen asks Amelia. 'People say you disappeared.'

'Yeah, lots of books have been written about me,' says Amelia.

Môrwen is silent for a moment. 'I can't read. I like picture books.'

'Didn't you go to school?'

'We have no schools in the Mesolithic. I learnt from my nan. Now I'm studying archaeologists,' says Môrwen.

'You mean archaeology?'

'No, archaeologists.'

'Well I'll just have to learn you to read, honey. I went to Columbia in 1919, majored in literature. I love books.'

Amelia writes the word 'Aeroplane' and draws *Friendship* in Môrwen's sketchbook.

'Amelia?'

'Honey?'

'Could we fly over the Tywi?'

'Well, *Friendship* was sold to the Colombian Air Force, and in the early '30s she disappeared. But hey, we're spirits, we can do anything!' and Amelia claps her hands, curses three times and they are flying over Carmarthen Bay in the spirited little seaplane. Môrwen sees the shape of an old ferry boat under the water and thinks, 'There must be a story there.' Amelia drops her near Llansteffan and vanishes, leaving a grin in the sky like a Cheshire cat.

The Ferryman's Coffin

Jacob the Ferryman owned a boat called *Rhondda* that carried people across the Tywi estuary, from Ferryside to Scott's Bay near Llansteffan. One of his regulars was a farmer called Ann Fawr, who had a son named Ben Bach. Ann was a widow, so Jacob helped her on the farm, and soon a romance developed.

One day in the sweltering heat of July, Ann thought she was going to die, so she sent Jacob to buy her a coffin. When the carpenter saw the measurements, he said there was no box big enough to accommodate her. It would have to be custom made, and that would be expensive.

Well, Ann wanted only the best coffin, so Jacob asked the carpenter to customise a solid oak box with brass handles. When Ann climbed in, it was too small, so Jacob blamed the carpenter for getting his measurements wrong and the carpenter blamed Ann for growing bigger since he measured her. So a few extra oak planks and a couple of iron supports were added, but it was still too small. Jacob accused the carpenter of incompetence and demanded a new coffin or his money back. The carpenter agreed to make another one provided he was paid in advance to cover the cost of felling a forest. They were about to come to blows when the carpenter agreed to delay payment until after the funeral, which according to Ann was getting closer by the day.

Years passed, and the feud continued until one day Jacob dropped down dead with the stress of it all, followed shortly afterwards by the carpenter, who had no one to argue with.

So Ann's little son Ben Bach, who was now Ben Fawr with four children of his own, took over the carpenter's job. He added more planks to his mother's coffin and said he would tie a rope around it after she was nailed in. Then one day Ben's children were playing hide and seek in the coffin when it collapsed. Ann was inconsolable without a box to be laid in to rest, and Ben was convinced his mother would probably outlive him and his children.

Then he had a brainwave. His mother was the perfect size to fit inside *Rhondda*, his father's old ferry boat. So when the day finally came, Ben placed Ann in the ferry, rowed out into the Tywi estuary, and sank it.

The Mock Mayor

During Fiesta Week in Llansteffan, the locals built a temporary stage in a clearing known as The Sticks for the Mock Mayor to make a speech. The custom had its roots in the irritation people felt at paying expensive tithes to the town Mayor, so they decided to elect their own. The contenders for Mock Mayor were hauled round the village in a gambo, making speeches that contained unachievable wild promises like those made in Westminster. One Mayor, Paddy Trench, promised to stock Carmarthen Bay with mermaids, but was defeated by the suffrage candidate long before women were allowed to vote in Wales.

Môrwen takes the ferry from Black Scar across the River Taf to Laugharne. The only other passengers are sheep from the Trefenty livestock sale. She arrives smelling of ewe poo and makes her way past the boathouse and ferryman's cottage up the steep tree-lined track to Brixtarw farm. There was another weird custom here, a sort of 'beating the bounds', which involved walking round the boundaries of Laugharne while someone from the crowd was held upside down or bent over as their names were recited at locations along the route such as 'Blinds Well'.

The Hell Dog

A woman was walking home alone from Laugharne at night when she stepped over a pothole by the roadside at Pant Madog and came face to face with a *Cŵn Annwn*, a Hell Dog. It pawed at the pothole and howled so horribly that the earth shook. Being a practical woman and unafraid of the dark, she glared at it until it vanished. She told her story in 1767 to Prophet Jones, who reasoned that someone had been murdered on this spot and the dog was trying to find their money, which had fallen into the pothole.

Years later a fisherman was walking home at night past the crossroads where the Laugharne town gibbet stood, when he saw a large white dog, growling and baring its teeth. The man ran, stumbled and fell, and the dog stood over him, drooling saliva. He closed his eyes and prepared to be torn to pieces, when the dog faded away.

During the Mesolithic, hyenas lived in nearby Coygan Cave, where food remains of mammoth, woolly rhino and brown bear, have been found.

The Ballad of the Long-Legged Bait

Môrwen walks along a muddy path, pausing only to chat with a herd of heifers on her way to St Martin's Churchyard, where Dylan and Caitlin are buried beneath a simple white cross. They lived in Laugharne for only four years, but Brown's Hotel is full of visitors searching for souvenirs of the poet, unaware his spirit is watching them from the bar and scribbling notes on the back of his cigarette packet.

Dylan spots Môrwen through the window. 'Mesolithic Girl, come in.'

'I have a name, it's Môr...'

'Landlady, a brown ale for this young lady, she's travelled a long way. Further than you could ever imagine,' shouts Dylan.

'You haven't paid last month's tab yet,' replies the landlady.

Môrwen grabs Dylan by the arm and ushers him away from Brown's before an argument begins. 'Can we go to your writing shed?'

'Yes, I have beer and Welsh cakes and questions to ask you.'

They stroll along 'The Walk' to a wooden shed by the waterside. It's not as tidy as the expensive Arts Council replica that toured round Wales on the hundredth anniversary of his birth. And it smells weird.

Dylan pours a bottle of brown ale into two glasses and gives one to Môrwen with an unbuttered Welsh cake. 'What's the Mesolithic like? How do you travel through time? Do you make a wish? Or have you got red shoes? Or a flying saucer? I'm writing a story about time passing in a small seaside town, seen through the blind eyes of a sea captain.'

Môrwen takes a breath. 'It only happens on this day when time moves in circles. You have to jump across when the lines meet. Have you heard of the Ballad of the Long-Legged Bait?'

'Is it about fishing for mermaids?' Dylan is hooked.

'Sort of. It's about my world. The graveyards below the sea, the miracles of fishes, the sun shipwrecked west on a pearl, old as water and plain as an eel, time bearing another son when there is nothing left of the sea but its sound. My people have been reeled in across time and space from a submerged land caught in the net of a thousand and one stories.'

Dylan frantically makes notes while a halo of cigarette smoke wraps round his head. 'A fisherman winds his reel with no more desire than a ghost, he catches a girl alive with hooks through her lips.'

Môrwen blows him a kiss, 'See you in Cei Newydd, lovely boy.'

18

Leekie Porridge
and Betty Foggy

Babs and the Flying Sweethearts

Môrwen is barefooting across Pendine Sands when a custom-made racing car called *Babs* hares past in an attempt to break the World Land Speed record in 1927. It doesn't end well, as the driver, John Godfrey Parry-Thomas of Wrexham, is killed when *Babs* overturns. She is now on display in the nearby Museum of Speed, and is commemorated by the hot rod races on the sands each year.

The Flying Sweethearts Amy Johnson and Jim Mollison, set off from Pendine to fly round the world in 1933 in a de Havilland Dragon called *Seafarer*. However, they crash-landed in Bridgeport, Connecticut, and became the first couple to fly the Atlantic, were entertained by President Roosevelt and met Amelia Earhart. Pendine has inspired strange things.

The Stonemason Bard

Tom Morris was born in a watermill in 1804 and lived with his wife Jenny in a cottage on the clifftop, where he sang ballads and played his cello to passers-by. He was a master stonemason who specialised in carving and engraving Snowdrop Marble, a grey limestone seen on many gravestones, memorials, fireplaces, churchyards and walls in South Pembrokeshire. His work is in Gloucester Cathedral, Tredegar House and St Mary's Church, Tenby.

Mesolithic Marros and Amroth

On Marros Sands are stumps of Mesolithic alder, oak and willow forest, alongside craters caused by exploded First World War mines, and the wooden outline of the schooner *Rover* wrecked here in a gale in 1886. On the beach at Amroth are groynes built from wooden sleepers from the Maenclochog Railway, which closed in 1949, and a row of cottages which were washed away in a storm in the 1930s. Môrwen picks up some plastic bottles and drops them in a bin, while a naturist paddles along the water's edge holding a bottle of Chardonnay.

Wally the Walrus

A huge walrus hauled out on the Tenby RNLI slipway in May 2021 and blocked the launch of the lifeboat. A crew member tried to move him with a broom while making noises with the air siren and waving his arms around. Eventually Wally flopped into the harbour, and set off on a grand tour of France, Spain and the Scilly Isles, where he sank a dinghy, tried to climb on to a fishing boat and made headlines in newspapers round Wales. After three months the orange giant swam round southern Ireland on his way home to Iceland.

Jane's Sublime Feelings

On a wrought-iron bench overlooking Wally's Slipway in Tenby Harbour sits a girl busily writing in a notebook. Môrwen peers over her shoulder. 'What you doing?'

'I am writing a diary of my travels in 1802. Listen: *Wales is not really somewhere to live; it is somewhere to have sublime feelings about, like a Gothic ruin or a mountain crag.*'

'I live here and I'm not a Gothic ruin. What's sublime?'

The girl closes her notebook and looks conspiratorial.

'The diary is merely a front to conceal that I am writing a novel about a girl who has run away to Wales to escape a man who wants to marry her. I confess it is somewhat autobiographical.'

'Is there a walrus in it?'

'My heroine is a matchmaker who refuses to marry.'

'Does she fall into a swamp?'

'The young man I am avoiding might be more interested in you. He prefers wild girls to the bookish.' And she scribbles out the words 'Sense and Sensibility', and writes 'The Marsh Girl and the Walrus', unaware that a once promising novelist has vanished from the literary and historical timeline.

Barti Ddu and Leekie Porridge

In the early 1800s, the captain of a cargo ship thought the Tenby harbourmaster's silver shoe buckles looked familiar. They were stolen when his ship was plundered by the Scots-American pirate John Paul Jones, who often hid on Caldey Island. In court, he discovered the harbourmaster had deserted from the navy to join John Paul Jones' crew, and his pirate name was Leekie Porridge, presumably due to the quality of his cooking. Porridge was found guilty of stealing the silver buckles and sentenced to serve on a British man o' war throughout the Napoleonic Wars.

Porridge wasn't as bloodthirsty as Welsh pirate captains like Henry Morgan of Llanrhymney, who plundered Spanish ships in the Caribbean, tortured his captives and ran slave plantations

in Jamaica, before drinking himself to death in 1688. When Howell Davies of Milford Haven was shot dead in 1719, a vacancy opened for another young Welshman, Bartholomew Roberts of Casnewydd Bach. Black Bart.

For three years Barti Ddu was the most feared pirate in the Caribbean. He employed a band to play his theme tune when he appeared on deck dressed in extravagant crimson clothes. Despite his showmanship, he never drank, banned gambling and prostitution on board, and ordered his crew to be in bed by eight o'clock. In January 1722 he was shot by a British navy officer and dumped overboard in his finery, so ending the short colourful reign of the first pirate to fly the skull and crossbones.

The Paddle Steamer

Môrwen has slipped on board the *Waverley*, the world's last seagoing paddle steamer, about to embark on a cruise from Tenby Harbour to Milford Haven. They paddle past Lydstep, where a 7,000-year-old skeleton of a pig was found with two broken flints in its neck, squashed beneath a fallen tree trunk on the beach surrounded by footprints. Môrwen loved that pig. It lives in Tenby Museum now, amongst Gwen John's paintings.

As the *Waverley* chugs past the old Mesolithic site of Manorbier, Môrwen waves to Gerald of Wales, who is wandering around the medieval fields behind his castle. At Stackpole there are dancing stones, at Castlemartin the military play wargames, and St Govan rings a silver bell to warn ships about the dangerous rocks near Bosherton. Govan's church is cut into the cliff face where he hid after being chased by pirates from Lundy, who stole the bell just before their ship sank after hitting the rocks. The bell washed up entombed in limestone, and still rings when struck. Oh – and count the steps on the way down the cliff to the church, and again on the way back. The numbers are never the same.

The *Waverley* sails past Sheep Island and Rat Island towards Thorn Island, where in 1878 the *Loch Shiel* ran aground with a cargo of whisky, beer and gunpowder on her way from Glasgow to Australia. Twenty-seven people were rescued by the Angle lifeboat, but the whisky mysteriously disappeared.

On 15th February 1996, the *Sea Empress* was on her way to the dystopian chimneys and domed storage tanks of Milford Haven's oil refineries, when she hit the rocks at the harbour entrance, spilling 73,000 tons of crude oil and causing an ecological tragedy that killed thousands of shearwaters, guillemots and puffins from Skomer, Skokholm and Grassholm.

Betty Foggy

A dark-haired baby was born in Pembroke Dock in 1805 to a Romany mother and a *gajo* man. She was named Betty but the day of her birth was misty so everyone called her Foggy. As she grew, she had a daughter of her own, who she raised alone. People told their children not to go near Foggy's ramshackle cottage, because she was weird and made potions with herbs and mushrooms, which she sold as charms and curses. Clearly, she was a witch.

On 21st July 1853, the people of Pembroke Dock dressed in their finest clothes to watch the launch of the ninety-gun wooden warship HMS *Caesar*. Foggy went too, but the dockyard police refused her entry as she was shabbily dressed. As she scurried away she muttered over her shoulder that everyone might as well go home as the ship would not be launched.

The inquiry into the failure of the launch concluded that the wood on the slipway was not greasy enough to allow the ship to slide smoothly into the water, though most people knew about Foggy's curse. The ship finally sailed on the evening tide on 7th August, when Foggy lifted her curse. She lived for another thirty years in her ramshackle old cottage, and Betty Foggy's Well is marked by a white cross on a rock at Trewent Point, for she is remembered while more 'important' men are forgotten.

Magic Roundabout

In 1980, rumours spread that another special ship was under construction in the Western Sunderland hangar in Pembroke Dock. The skilled shipbuilders were sworn to secrecy, referring to it by a codename, 'Magic Roundabout'. The schoolchildren knew the dockyard had already constructed one of the sets for Stanley Kubrick's *2001: A Space Odyssey*, and soon the secret was out: the ship was the Millennium Falcon, Han and Chewie's old spaceship and star of *The Empire Strikes Back*.

Haggar's Electric Picture Palace

The early twentieth-century Welsh movie pioneer William Haggar showed his films such as *A Message from the Sea* and *The Salmon Poachers* at his own moving-picture palace on Main Street in Pembroke. *The Maid of Cefn Yddfa* was based on a Welsh folk tale, while *The Life of Charles Peace*, the murderer, was filmed mostly in Pembroke Dock and featured the Haggar family in starring roles. William's grandson Len kept the cinema running until 1982, and when he showed *Lawrence of Arabia*, Len's daughter Dinah sold more ice cream than ever thanks to the desert heat.

The Dragon of the Wogan

Around 11,000 years ago, Môrwen's nomadic ancestors lived in the Wogan, the cave beneath Pembroke Castle where bones of mammoth, wild horse and reindeer have been found alongside stone tools. It was walled off in the early thirteenth century, but was eventually connected to the castle by a spiral staircase. One day some boys climbed down to the cave, where they found a fearsome dragon that killed their dog. It turned out to be a small crocodile left there by a sailor who thought it would make a nice pet.

Tōgō's Ginkgo Tree

Lieutenant Tōgō Heihachirō of the Japanese Navy lived in Pembroke Dock during the construction of the battleship *Hiei*, which was launched at Jacob's Pill on 9th June 1877. When he returned to Japan, Lt Togo sent a ginkgo tree to Pembroke Dock with the message, 'Please plant this tree in the garden of my lodging house in appreciation of the kindnesses shown me during my stay.' The tree is still there in the Master Shipwright's Garden.

Staff from the National Botanic Gardens of Wales took cuttings and grew saplings, which were sent to Hiroshima City Botanic Gardens, and the first Welsh gingko was planted in Kure Naval base on 1st July 2020. All thanks to Lt Tōgō of Kagoshima, who trained to be a samurai warrior, became Admiral of the Fleet during the Boxer Rebellion and whose own garden was recreated at the National Museum of the Pacific War in Fredericksburg, Texas, as 'The Japanese Garden of Peace'.

The Quaker Whalers

Fifteen Quaker families from Nantucket and Martha's Vineyard, led by Samuel Starbuck and Timothy Folger, arrived in thirteen ships in Milford Haven in 1792 to develop a whaling fleet. The ships hunted the Southern oceans in perilous seas and were away for long periods, so the trade in whale oil lasted only thirty years. However, they left their mark on the town with Starbuck Road, the Quaker Meeting House and the Museum, which used to be the whale oil store, and Herman Melville used the name Starbuck for the first mate of the Pequod in *Moby Dick*.

How to Choose a Husband

The Llangwm Oysterwomen sold herring from door to door dressed in red flannel petticoats and black-brimmed felt hats tied with white scarves. Dolly the Bridge, who lived by the Guildford Bridge, inspired artists to paint her, and earned a living selling coloured postcards of herself as an oysterwoman. In 1863 she chose her own husband, William Palmer, and raised ten children. When she died in 1932 aged 90, she had twenty-six grandchildren and thirteen great grandchildren.

Choosing husbands was a thing in Milford Haven, especially on the *Teir nos Ysprydion*, the three spirit nights. One girl left a plate of cheese on toast and a glass of beer on her table, then shovelled coal on the fire, took off her clothes, washed her pants in a bucket of cold spring water, pulled them out of the tub upside down and to the left, and hung them on the back of a chair to dry. She climbed into bed, wrapped herself in a warm quilt and listened to the silence. She heard a noise and peeped through the keyhole to see an apparition of her future husband. Mostly this worked, until a girl in Tenby saw a huge black hairy monster with fiery eyes stuffing itself with cheese on toast. She stayed single.

19

Skomer Oddy
and the Smalls

ABERDAUGLEDDAU/MILFORD HAVEN–YNYS SGOMER/
SKOMER ISLAND–BAE SANT FFRAID/ST BRIDE'S BAY

The Merman

In 1782, farmer Henry Reynolds from Castlemartin saw a wild, white-skinned youth of about 18 bathing off the north coast of Milford Haven. The lad watched the birds flying overhead through fierce wild eyes separated by a Roman nose. Mr Reynolds noticed a brown tail below the water, waving like seaweed, so he fetched his friends but the merman had vanished.

Skomer Oddy

Amser maeth yn ôl, a long time ago,

Giants lived on the mountaintops in Pembrokeshire, and some were mean. Very mean. They trashed houses, felled forests and had wars with the coastal people. The giant who saved Milford Haven wasn't mean, though. He was Skomer Oddy.

Skomer was strong. He once carried a pile of massive bluestones on his shoulders from the Preselis to Milford Haven and tied them to a wooden raft. He dragged it through the coastal waters and overland to Wiltshire, where he built an enormous stone henge.

Two sea serpents were fighting off Milford Haven. They twisted their bodies around each other, broke fishing nets, stirred up the sea and destroyed the precious mearl beds. Soon the Haven was full of brown sludge – not the usual oil or sewage, but mud. The mermaids were seriously annoyed. Their houses were covered in mud. The sea serpents had to go.

So the mermaids went to Foel Cwmcerwyn, the highest point in the Preseli Hills, and followed the snoring to Skomer's hidden cave, where he woke up every hundred years to help people before returning to his beauty sleep. The cave was dark and dank and smelly, because the giant hadn't washed in almost a hundred years. They clambered over him and sang in his ears:

Skomer Oddy! Skomer Oddy!
Big head and big body!
Help us, help us! Help us, help us!
Get the mud back in the sea.

Skomer shook off the mermaids, turned over, and went back to sleep.

Skomer Oddy! Skomer Oddy!
Shaggy hair and smelly body!
It's time to wake and help us
Get the mud back in the sea.

Skomer mumbled, 'It's too early! Twirly. Like a pig's tail.'

The mermaids told him it was time to wake up. He'd been asleep for nearly a hundred years and serpents were trashing the estuary.

Skomer grunted, 'I cleared up the oil after the *Sea Empress*. Go 'way.'

So the mermaids hauled him out of his nice warm bed, wrapped his cloak around his shoulders, clamped a staff the size of a tree trunk in his hand, kissed his forehead and shoved him through the cave entrance. He took a deep breath, filled his pockets with mermaids and walked with giant strides to muddy Milford Haven.

The serpents hid in the mud when they saw him coming, but Skomer dragged them into the deep ocean, threw them over Skomer Island towards Ireland, and they were never seen again.

That's why Milford Haven is the only site in Wales where the rare, slow-growing red seaweed Mearl grows.

As Skomer cleans the mud off the beaches, one of the mermaids kisses his knuckles. He scoops Môrwen out of the water with his giant hand, stuffs her in his pocket and strides around St Ann's Head, away from the petrochemical Haven towards the enchanted green islands.

The Green Fields of Enchantment

Off the west coast of south-west Pembrokeshire were green islands that disappeared in front of your eyes when approached, not by slowly sinking or drifting into the mist, but suddenly. The otherworldly folk who lived there visited the markets at Milford Haven through a secret subterranean tunnel. They never spoke to the traders, but always left the exact money. Sometimes they were invisible, though one butcher saw them. They were Môrwen's people. Plant Rhys Ddwfn. The island was Skomer's.

Skomer Voles, Seals and Puffins

Skomer strides across Jack Sound towards the island named after him, through water so crystal clear he can see the seals swimming on the seabed. As he sits down on the island, an army of voles crawl up his legs and tickle his body until he rolls around, giggling uncontrollably and squashing the autumn squill. Skomer Voles are unique to the island. They arrived long ago on a supply boat, and now there are 20,000 of them, but visitors are too distracted by the cute puffins to see them.

Môrwen climbs out of Skomer's pocket and follows the field boundaries, stone walls and earth banks around the Wick, past Pig Stone and Mew Stone, towards the burnt mounds where Mesolithic people built nomadic huts with stone and driftwood. Her foot slips down a burrow, and when she pulls it out, a puffin chick looks up.

'You trod on my beak!'

'Well, you shouldn't dig a burrow in the middle of the path, silly bird.'

'And you shouldn't step on an endangered species, rude girl.'

The puffin opens its beak, sticks its tongue out and makes a raspberry.

Môrwen is about to answer when Skomer rescues her from an argument with the puffin, stuffs her back in his pocket and wades through a cloud of storm petrels towards Skokholm Island.

Alice and the Wild Man of Skokholm

A little girl called Ann is playing chess with a lobster and a hawk outside the farmhouse on Skokholm Island. Her father, the writer Ronald Lockley, was warden of the island when he wrote *The Private Life of the Rabbit*, which inspired Richard Adams to write *Watership Down* and led to the film that gave nightmares to a generation of children. The Skokholm bunnies became archaeologists in 2021, when they dug a burrow and unearthed a 9,000-year-old Mesolithic bevelled pebble.

When the schooner *Alice Williams* was wrecked on the island in 1928, Lockley bought the salvaged timber for £5 to rebuild his farmhouse. He placed Alice's figurehead on the cliff overlooking South Haven, where she stayed until she lost an arm in the storms. She was renovated by an artist from Dale, so Alice still lives on the island of rabbit burrows.

One of the island farmers was Jack the Bulldog, a short strong man who owned a herd of cows, a hundred chickens, four pigs, many rabbits, a few horses and a goat who could speak, like all goats in west Wales. He grew root veg and hay, fished for lobsters and sold eggs and butter on the mainland, where he quoted the Bible and got into fights. The Bulldog employed a lad called Edward Pearce, who had a thick beard, long hair and bulging eyes, and once ate thirty boiled eggs in an evening after he was challenged in the pub. Hence his nickname, the Wild Man of Skokholm.

Grassholm and Gwales

Grassholm is uninhabited – apart from 39,000 pairs of gannets who cover almost half the island with nests built partly of plastic waste found floating on the sea. The island reeks of their guano, though an eleventh-century bronze Hiberno-Norse sword guard was found here, suggesting Scandinavian migrants with little sense of smell.

King Bran lived here for eighty years. After the wars with Ireland and the death of his sister Branwen, he ordered his noble head be cut off before the poison from a foot wound reached his brain. His friends brought him here to Grassholm, then called Gwales, where he lived under an enchantment, talking, eating and drinking, until a door facing Cornwall opened and the spell shattered. He is buried in the Tower of London, guarded by ravens. Bran means Raven in Welsh.

Môrwen senses Branwen's spirit. She can't escape war and death. Grassholm was used for bombing practice during the Second World War . Though no one stays dead for long in Wales, especially on Calan Gaeaf.

The Smalls Lighthouse Tragedy

Twenty miles further out to sea is another island, the western-most rock in Wales. The lighthouse on The Smalls was built in 1776 following designs made by a musical instrument maker from Liverpool, to prevent ships being wrecked on the Barrels and Hat reef. In 1801, the lightkeepers were Thomas Griffith, a labourer, and Thomas Howell, a cooper. They disliked each other intensely despite both being family men and neighbours from Solva. Their feud lasted until Griffith fell ill and died, leaving Howell worried he would be accused of murder. Howell used his carpentry skills to build Griffith a wooden coffin, which he kept in the lighthouse until the stench of the rotting body forced him to hang it outside the window. The storms pummelled it until Griffith's arm hung loose and swayed like the ticking hand of a

clock, as if accusing a guilty man. After weeks of bad weather, a boat from Milford Haven reached The Smalls to discover a dead man and a mad man. They later became Willem Dafoe and Robert Pattinson in the movie *The Lighthouse*.

Mesolithic Nab Head

The cliffs at Nab Head are surrounded by sea, but when Mesolithic families lived here 10,000 years ago, the coast was 6km away to the west. Flints, microliths, drill bits, and scrapers have been found, along with 700 perforated shale beads, which Môrwen strung together to make necklaces. Her ancestors hunted, counted stars, listened to the sea telling stories, and followed the deer herds through the submerged forest at Newgale, which was exposed by a violent storm in the early 1170s, when tree trunks rose from the sea like wind turbines.

Stranger Things in St Bride's Bay

A farmer once caught a chestnut *Ceffyl Dŵr* at Whitesands and fastened it to his plough. So it dragged both farmer and plough into the sea, just like the horse on the banks of the Tywi. One dark evening, a witch flew across the road towards the Two Foxes in Marloes and disappeared into the roof. Marloes is known for strange stories, but even stranger things were witnessed in Broad Haven in 1977.

On 4th February, a yellow cigar-shaped object landed in a field near the primary school, and when the headmaster asked the children to draw what they had seen, their sketches were almost identical. In March, a tall man with high cheekbones was seen dressed in a one-piece suit, followed in April by a hovering egg-shaped object with an orange-red light. Then a girl found a small alien sat on her bedroom windowsill, and a faceless pointy-headed creature with long arms and legs appeared in Little Haven. An upside-down saucer left scorch marks in the grass, and a glowing object chased a car home to a farm overlooking Stack Rocks, where the driver witnessed cattle herded by a man in a silver suit. Some said these strange things were from RAF Brawdy, although the children, now grown up, know what they saw.

Callican's Donkeys

In the 1880s, Edward Callican of Solva kept twenty donkeys who carried culm and wood from Newgale beach for the poor folk to burn on their fires. He charged two pence a sack, or a penny if you provided your own donkey. And he carried lime from the stench and heat of the twelve circular kilns by the harbour for farmers to spread on their fields.

Callican was one of the last to go 'Hunting the Wren' at Twelfthtide, Old Christmas, 6th January. He carried a little wooden wren cottage decorated with ribbons from door to door, while he sang rhymes in a thin quavering voice intelligible only to dogs:

Joy, health, love and peace be all here in this place
By your leave we will sing concerning our King
Our King is well dressed in the silks of the best
In ribbons so rare no king can compare.

Also in the 1880s, H.W. Evans made a model of a *Mari Lwyd* from Solva, now in the National Museum of Wales. It was made from a square of tabby cotton folded into a triangle to make a pointy face with mouse whiskers, ears cut from dark blue stiffened linen padded with cotton wool, and a face painted with brown ink. And this still wasn't the strangest thing.

The Giant Boar-Man

Young Culhwch wanted to win the hand of Olwen, so her father, the giant Ysbaddaden Bencawr, set him the task of cutting the comb from between the ears of Twrch Trwyth, a king who had been transformed into a terrifying half-boar, half-man. After a chase across Ireland, Twrch Trwyth swam to Porthclais, and left terrible slaughter across Pembrokeshire. Only when King Arthur's men cut the comb from the boar-man in the Cornish sea did the bloodshed end.

Skomer places Môrwen in a rockpool amongst a colony of 100-year-old pink sea fans, tiny creatures who share communal hard skeletons. She kisses his gnarled knuckles and he tramps off to his cave on the Preselis, snuggles into his duvet of a thousand and one fleeces, and sleeps until he is needed to clear up the next oil spill, lifeboat rescue or boar-man rampage.

20

The Old Fibber and the Last Invasion

The Ghost Island

Gruffydd ap Einion, fisherman and market trader, stared out to sea from St Davids Churchyard and saw a ghost island, but when he ran to his boat it vanished. He thought it must be a mirage, until an old woman explained he was standing next to the one patch of herbs that allowed people to see the land of Plant Rhys Ddwfn, the Welsh utopia. So next day he returned to the churchyard, and the moment he saw the island, he dug up the herb with a ball of soil round its roots and planted it in his boat. With the island visible, he rowed towards it and landed in a cove, where he was welcomed by Rhys. Every evening he visited until one day, he never returned. People said he had gone to live with the fairies, where he belonged.

A fisherman was watching the sunset from his boat at St Davids, when he saw a little man climb up the anchor chain and clamber on board. He shook his fist and ranted '*Pobl bach! Pobl bach!*' because the anchor had crashed through his roof, landed on his dinner table and left a big lump on his head. You see, *Pobl bach* live beneath the sea, too. Indeed, in 1858 Captain Daniel Huws saw an underwater town full of mermaids off Trefin.

How St Justinian Made Ramsey Island

St Justinian was a Breton nobleman who lived at St Davids, but the monks disliked him so he walked out to sea along a causeway at Porthstinian and broke the rocks behind him to stop them following. The broken rocks are Ramsey Island, the Bitches, and the Bishops and Clerks.

Justinian lived as a hermit on Ramsey with only the sound of the sea and seals for company, until St David ordered him to return to the cathedral. This time the monks chopped off his head, so he picked it up, returned to Ramsey and buried himself amongst the seals. Though when he was sanctified, his bones were reburied in St Davids graveyard, where he never wanted to be.

In St Davids Cathedral is a misericord with a carving of St Govan being seasick in a small boat, while three men smile and pat him on the back.

The Beekeeper

An Irish monk called Modhomhnóg was the beekeeper who supplied St David with honey and wax candles. One day he left the cathedral to catch the boat home to Ireland, so the bees followed him. He returned the bees to their straw skeps, but they followed him again. Three times this happened until St David told him to take them with him. So Modhomhnóg built new hives in a Welsh garden at Bremore, near Balbriggan, and that's how honeybees were introduced to Ireland.

Mesolithic Whitesands Bay

A submerged forest lies in the peat beneath Whitesands Bay where Mesolithic flints, deer bones, auroch horns and a bear jaw have been found. Old Teithi Hen lived here until the floodwaters forced him to ride to dry ground, where he was given refuge by King Arthur himself.

The giant Samson built a cromlech a little further along the coast near Abercastell 5,000 years ago, with seven uprights and a huge capstone that he lifted into place with his little finger.

Carregwastad Mermaid

At Carregwastad in the early 1700s, one of Môrwen's sister mermaids was caught by some fishermen from Pencaer. They held her captive in Mr Mortimor's wine cellar at Trehowel farm and fed her on fish soup with sweet milk to make it palatable. She told the fishermen to release her or they would be horribly cursed, and as she swam away, she promised to warn them of impending storms, plagues or invasions.

The Last Invasion in 1797

Môrwen hauls out on the rocks at Carregwasted, combs her tangled, seaweedy hair with a crab claw and stares at her reflection in a green glass float attached to a lobsterpot. She finds a bottle of Madeira, hidden between the rocks by the mermaids after they salvaged it from the wreck of the Portuguese wine ship *Friends*, which sank here in December 1796. As she takes a swig, four ships with ghostly white sails drift through the twilight from Strumble Head, a flotilla of small boats row towards the cliff, and an army of soldiers crawl around the rocks like confused ants.

Môrwen stares at the bottle of Madeira and realises she has slipped through time to 1797 and is in the middle of the Last Invasion. She remembers the promise made by her Carregwastad sister, so she sings a warning to the Pencaer fishermen while her other sisters drag one of the boats into the depths. The fishermen alert the Fishguard Infantry to take up arms. The Infantry are only a couple of hundred local men with little combat experience and bullets made with lead from St Davids Cathedral roof. The invaders are from *La Légion Noire*,

a patchwork army of criminals, prisoners, deserters, activists and innocents who tried to invade England in December 1796 but were blown off course.

Môrwen wraps herself in a blue shawl and follows La Légion to Trehowel farm, where they drink the contents of Mr Mortimor's wine cellar. One troop breaks into Llanwnda Church and lights a bonfire with Bibles and pews, while the rest go pillaging and plundering. It all feels a bit *Whisky Galore* until Môrwen finds herself consoling a farmgirl whose husband was shot while trying to stop his wife being raped. This is a mess.

Môrwen hurries into town to warn the shoemaker Jemima Niclas, nicknamed Jemima Fawr, who rams her black floppy hat on her head, wraps herself in a red shawl, rounds up a group of soldiers with her pitchfork and locks them in St Mary's Church. Jemima realises the Infantry are outnumbered, so she calls at every house until she raises an army of 400 women. Wearing blue and red woollen shawls to look like soldiers, they march up the hill above the Parrog with shovels and brushes over their shoulders.

Jemima's women tramp round and round the hill to give the impression the Welsh army is bigger than it is, but *La Légion* aren't fooled. They know the old stories of women who dressed as soldiers to scare invading fleets of ships. What strikes fear into them is that some of these ferocious Fishguard women have tails sticking out from beneath their skirts, and several boats have already been sunk by the local mermaids. So they surrender at the Royal Oak, and are locked up in Ha'rford Jail before being deported, so ending the two and a half days' war.

Jemima's tombstone reads, 'The Welsh Heroine who boldly marched to meet the French invaders who landed on our shores in February 1797. She died in Main Street July 1832 aged 82 years.'

Wales hasn't been invaded since, thanks to the many stories of the General of the Red Army. And a few hundred mermaids armed with bottles of Madeira.

The Great White Whale

One dark night in 1954, a lad was walking along the coast at Strumble Head when he saw a ghostly white shape drifting round the cliff. It looked like a monster sperm whale, but it couldn't be – except the whale was very real.

Well, sort of. It weighed 20 tons, was 20m long and 6m high, with a mechanical steel frame as the middle section, but with no head or tail. It was one of two prop whales from the movie *Moby Dick* that John Huston was filming 5 miles out at sea from Fishguard Harbour. The script by Ray Bradbury had been tweaked by a rising young Welsh writer, Roald Dahl, and included a character called Starbuck named after the Milford Haven whalers.

Moby Dick's co-star, Hollywood heartthrob Gregory Peck, was sat on the half-whale's back in his role as obsessive Ahab when a rope broke. Free from the tugboat that was hauling it, the metal whale drifted into the fog, rocking from side to side in the strong waves. Peck lost his footing on the wet surface and landed in the cold sea. He wrote in his autobiography that he could see the headlines: 'Movie Actor Lost on Rubber Whale', as the star of *To Kill a Mockingbird* followed Ahab to a watery grave. Fortunately he was rescued by the coastguard and returned to his lodgings at the vicarage, and the unrequited affections of the vicar's daughter. The whale was never seen again.

The Flyer and the Fibber

At 6 a.m. on 22nd April 1912, Denys Corbett-Wilson – war hero, jockey, racing driver, and playboy son of a rich barrister – took off in a Blériot monoplane from a field overlooking the ferry terminal in Goodwick, and 100 minutes later landed in a field in Enniscorthy, having completed the first flight across the Irish Sea. However, Corbett-Wilson was not the first to fly across the land of Plant Rhys Ddwfn. That honour belonged to an old fisherman and fibber from Wdig.

Shemi Wâd loved to tell stories over a pint to the regulars at the Rose and Crown, who always suspected he was fibbing them. Shemi's favourite things were fishing and sleeping, often both at the same time, and he had half-shares in a fishing boat with his mate Dai Reynolds. They caught mackerel, lobsters and crabs off the Cow and Calf Rocks, and split the catch equally, though Shemi said to be fair he'd have the fish on the inside of the net and Dai could have those on the outside.

They once caught a giant herring, far too big to fit in the boat, so they hauled it back to Goodwick in the net. When they cut it open, old Jonah himself hopped out, and don't laugh, because he wasn't the first Bible character to turn up in Fishguard. Years before, Jesus called at a cottage and asked a woman if he could bless her family. Well, she had seventeen children and thought Jesus might disapprove of so much sex, so she hid ten of them in the woods. Jesus blessed the remaining seven and left. She went to fetch her other ten, but couldn't find them. They became the first fairies, *y bobl bach*. Old Shemi's ancestors.

After a pint, Shemi took his rod and flies and went fishing off the Parrog. He caught the biggest sewin ever, and was thinking about supper when a heron swooped down, swallowed the fish whole and flew away.

'No, man!' he shouted, still holding the fishing rod, for he wasn't going to let that gangly-legg'd bird have his supper. The heron flew into the clouds, hauling Shemi after it.

The heron saw a big rock in the middle of the sea, and landed with a thump, so hard that the sewin popped out of its beak. Shemi grabbed it, licked his lips at the thought of fish supper, and realised he was stranded on a desert island.

Just then, a giant crab passed by, so he jumped on its back and sailed to Goodwick, leaving behind the imprint of his boots on the rock. When they reached the Parrog, the poor old crab died from exhaustion and everyone had crabmeat pie at the Rose and Crown, to celebrate the day Shemi Wâd flew to Ireland.

Shemi's gravestone at Rhos-y-Caerau, Pencaer, is overgrown with brambles, but it reads '*Cyfaill i bawb a hoff gan bawb*', 'A friend to all and loved by all.'

And that's no fib.

The Cottage on the Rock by the Sea

At the beginning of summer, a father and his sick daughter arrived at the newly opened Goodwick Railway Station and rented a cottage in Lower Fishguard. Each day the girl sat in the doorway to breathe the sea air, until slowly she gained strength. She fished for crabs on the quay and wished she could go to sea in one of the wooden fishing boats. One day in August, her father hired a boat and rowed her to the next beach, where they built sand castles. Each day they rowed to a different beach but she loved Pwllgwaelod best. Her father decided to build a house there, so the family could stay every summer until she was well again. They chose a rock with a view, her father brought a trowel and mortar, laid a foundation of stones from the beach, made walls with square windows, used flat pebbles for the roof, and within a week the cottage on the rock by the sea was complete.

The girl and her father never came back. The cottage has eroded and been rebuilt, a mysterious memory of how the sea brought life to a sick girl.

21

The Mermaid and the Swan Girl

Weird Tales from the Teifi

Where the Teifi flows into the sea, a skeleton swims underwater as if beneath glass, but when the ripples break the surface, it vanishes. Another skeleton rows a boat below the bridge while the wind whistles eerily through its ribs. A girl is lifted from the sea by a huge bat and transforms into a witch. Two brothers at Y Ferwig use conjuring and arsenic to cure cancer. A girl called Siwan stands on the clifftop on stormy nights as a mermaid. Upriver at Cilwendeg is a grotto built of seashells in the 1820s. Search for 'Cardigan folklore', you get photos of Taylor Swift wearing a cardy.

In the late 1780s, Pergrin, a fisherman from St Dogmaels, caught a mermaid in his nets at the mouth of the then unpolluted Teifi. She told him to release her, and promised to howl three times if he was in danger, so he threw her back into the sea. Time passed, until one cloudless day, Pergrin was out in his boat when he heard howling. He hauled in his nets and sailed for shore as a storm gathered. He reached safety, though twenty-seven fishermen drowned that day, 1st October 1789. A wooden statue remembers the unnamed mermaid near the Coast Path in St Dogs.

Shame nobody heard her before Christmas 2023, when Welsh Water admitted illegally spilling untreated sewage from Cardigan Wastewater Treatment Works into the Teifi estuary and thirty rivers across the country, which showed little respect to the Cilgerran coracle fishermen or the water buffalo who were introduced in 2017 to graze the Teifi Marshes.

The Storybus

Môrwen catches the T5 bus from Sgwar Alban towards Cei Newydd. This is a storybus. One man talks about the local badgers and gives her a Siani Chickens lapel badge, while another hands out flyers about veganism, and a couple of ladies do a pole dance while ringing the bell.

Cardigan Island: Thirty pairs of puffins bred here, until the liner *Herefordshire* was wrecked in 1934 and the ship's rats scurried ashore and wiped them out.

Mwnt: Human bones have been unearthed beneath the hill, where wild games were played on *Sul Coch y Mwnt*, Red Sunday.

Aberporth: Carreg y Môrwynion, the Mermaids' Rock, where a group of girls drowned after becoming trapped by high tides, and there is a wooden carving of a mermaid in the Ship pub.

Cranogwen, Master Mariner: Sarah Jane Rees of Llangrannog was taught navigation by her father John Rees, captain of a ketch that transported cargo around the coast. She qualified as a sea captain at 21 and won a poetry prize at the National Eisteddfod under the Bardic name Cranogwen, aged 26. When she became a schoolmistress, her students were known as Cranogwen's Captains. Like Myra Evans, she is a Ceredigion heroine.

Carreg Bica: A rock in the sea north of Llangrannog was spat out by a giant with toothache.

Synod Inn: Siôn Cwilt the smuggler was a wild-haired giant man who wore a patchwork quilt for a coat and lived in a *tŷ unnos* near Synod Inn. He supplied Sir Herbert Lloyd MP, the wickedest man in Ceredigion, with brandy, and in return

Sir Herbert turned a blind eye to the smiles on people's faces. Until one night in 1797, a ship from Roscoff pulled into the bay at Cwmtydu with a cargo of brandy that was stowed in caves along the coast. Smuggler Daniel Ifan was hanged in Llangrannog, sparking protests in Cei Newydd as the coast swarmed with troops and excise men, but the liquor had vanished and so had Siôn Cwilt. Yet the smuggler is still so popular the local school is named after him, Ysgol Bro Siôn Cwilt.

The Invisible Fishergirl

Môrwen gets off the bus in Cei Newydd next to the house where Myra Evans lived in the late 1800s. The fishergirl from Llanina has sold a lobster to the landlady of a pub decorated with fishing nets and portholes. Two diners are studying the menu, wondering whether to go back to their camper van and cook the cheap food brought from their supermarket back home. They decide to treat themselves to local lobster, unaware that the fishergirl who caught their dinner is in Costcutter spending her money on reduced sausage rolls and frozen oven chips for her mam's dinner.

The Commando who
Tried to Shoot Dylan Thomas

Two men wave to Môrwen as they load beer bottles into a pram outside the Black Lion. One is Myra Evans' son Aneurin, now a giant of a man dressed in a sombrero and a tent-sized flowery shirt, who lives up to his nickname 'King Kong'. The other is Dylan, his mischievous schoolmate from Swansea and Laugharne, who shows Môrwen a scrap of paper.

'I've written a poem about your Mesolithic people and the graveyards under the sea. I call it "Ballad of the Long-Legged Bait".'

'Are there Miracles of Fishes?'

'Of course!'

They link arms and follow the cats' eyes along the road to Llanina, looking for all the world like Dorothy, Scarecrow and Cowardly Lion.

Dylan moved to Llanina from London in September 1944, with Caitlin and their two children, to help his childhood friend Vera Williams and her husband William, a commando captain who returned shellshocked from the war. By a fairytale coincidence, in the same month of the same year, Myra, Aneurin and youngest daughter Iola moved to the little blue house down the road in Gilfachreda. The neighbours from Swansea were together again.

Dylan recites his poem as they walk to Majoda, his ramshackle bungalow on the clifftop overlooking the mermaids of Cardigan Bay. He unloads the beer bottles into the *tŷ bach* while Aneurin and Môrwen walk towards Myra's house. They hear gunshots, and turn to see William standing outside Majoda holding a machine gun and a hand grenade. The walls are peppered with bullet holes.

Myra hears the commotion and runs down the lane to find the Thomas family shaken but unhurt. William had left the Black Lion following an altercation about the poet's friendship with Vera. He was arrested and charged with attempted murder, but found not guilty. The Thomases left Ceredigion shortly afterwards, and never returned.

Môrwen's Story

Myra links arms with Môrwen as they pass Llanina Church and walk along Afon Gido to the beach. They spread a chequered tea towel with one of Dylan's beers, two glasses and the remains of the octopus salad. Môrwen draws Branwen, Arianrhod, Dwynwen, Sabrina and Aerfen in her sketchbook, while Myra draws Dylan smoking a cigarette, and adds, '*Gofynnodd i mi ddysgu Cymraeg iddo.*' 'He asked me to teach him Welsh.' They laugh uncontrollably.

Môrwen takes a drink and says, 'I'll tell you a story I've never told anyone before.'

'One Nos Calan Gaeaf, I stepped through the veil into 1859 and watched a young fisherman checking his lobsterpots at Ogof Deupen. He had dark curly black hair and sang about a mermaid with spun gold hair. I was enchanted. You know, mermaids and fishermen?

'I asked him to teach me his song, so he rowed closer and sang again, and I flushed pink as a sunset. That had never happened before. I was annoyed.

'His name was Rhysyn and he lived at a smallholding called Tangeulan with his mother, Nidan. He told me he liked wild fish girls, but he was to be married the following day to Lowri the maid of Plas Llanina, a Valleys girl with hair of spun gold.

'I twisted my tangled seaweedy hair between my fingers, thrashed the water with my tail and cried salt tears. I stared him in the eyes and told him if there was a wedding, the sea would take his home. And I dived beneath the waves.

'Calan Gaeaf dawned bright, but as Rhysyn entered the church, the skies turned grey and the storm blew through the church doors and swept the congregation out to sea. Horses and sheep drowned that day, and dolphins swam where people once walked.

'I wrapped my arms around Rhysyn and held him until his legs became a tail. He shared my coral bed amongst the shipwrecks where the handmill was still grinding out beer, salt and mermaids, but he never looked at me. He stared at the reflection of the moon above the sea. One day he vanished. I never saw him again.'

Yole

Myra fumbles in her bag and gives Môrwen a copy of her book of local fairy tales, *Casgliad o Chwedlau Newydd*. Môrwen gives Myra her travel sketchbook, the two women kiss, and arrange to meet on the beach same time next year.

Môrwen strokes the texture of the dark green book cover and opens it with her fingertips. Her own story is there. On page one. How did Myra know?

Môrwen has drunk plenty of beer today, so maybe she has imagined it all? No, that would be far too corny.

She looks up to see a swan flying towards her. It removes its wings and feathers and out steps a girl with white hair and a *kircher*, a bonnet.

'I know you,' says Môrwen. 'You're Swan girl?'

'I'm *Yole*,' says the girl,. 'It means "old" in my language.'

'I'm named after the sea. Y Môr. Friends call me "Marsh Girl".'

'I've flown over flooded river valleys from Ireland. The water is rising.'

Yole isn't speaking Welsh, Irish or English, but Môrwen understands the shared language of girls who can transform into birds. There's no need to rationalise this. Enslaved Africans flew home to Nigeria from the Outer Banks in the 1800s. Many people witnessed it and their observations were recorded in print. People flew. It happened. Magic is best left unexplained.

'How old are you?' asks Môrwen.

'Maybe 7,000,000 beats of a swan's wing?' says Yole.

'I'm a flint in a plastic bag. Do you know the Children of Lir?'

'My sisters Aodh, Fionnghuala, Fiachra and Conn were cursed by our stepmother Aoife to travel across Ireland as *Eaaí*. Swans. When the enchantment wore off, they transformed into old people who spoke the language of birds.'

Môrwen smiles. 'One of my ancestors lived on Llyn Glasfryn. She told herself fairytales as she transformed into a swan. Another lived on Gower and two more on Barry Island. They spoke the language of swans.'

'The coastal farmers around Rosslare and Kilmore speak my language,' says Yole, 'my first word was *mulke* when my mother gave me her breast. I played *keek*, peepo, with my *vroene*, friends, and I called the family goat, *gurthe*. If I felt a little *drazed*, I said everything was *quare* bad or *quare* good or just *quare*. My parents called me their *quare* girl, and said I was away with the fairies, but not *tylwyth teg* or *pobl bach*. I called them *ammache*.'

Môrwen and Yole say that last word together and lie back on the sand in fits of giggles.

They are so different they are exactly alike.

They stare at their reflections in the sea, and two swans peer back at them.

Time and space have merged in that fraction of a microsecond of life that passed while you read this word. Just now, in a minute.

Swan Girls Fly Home

These two sisters are from a time when humans shared their ancestry with the sea. Time and space mean nothing when you're mythological. They know no borders. They need no passport to swim. They smell the air, raise their arms and fly as birds.

The people on the beach look up as two swans fly from the coastal floods. Yole flies to the Northlands. In Denmark she dies in childbirth, and is buried wearing a necklace of snail shells, her pelvis and cheeks red with ochre blown from a bowl, a flint knife at her hip as if she is a man, with her baby wrapped in a swan's wing. Death is only a small moment of her life.

Môrwen is weary of twenty-first-century noise. In a wingbeat, she skims over the water to her home in the land of Plant Rhys Ddwfn. She will survive the floodwaters, for she can swim as a mermaid and fly as a swan.

She has travelled round the coast in a day, ridden a water horse and a she-wolf, met the wise old toad, changeling children, a mammoth, sea dragons, river spirits, lifeboat crews, cockle gatherers, bonesetters, migrants, mapmakers, travellers, radicals, not to mention Branwen ferch Llyr, Skomer the giant, the King of Bardsey and King Elfys of Memphis and Preseli, and the archaeologists who are studying her own Mesolithic people. And of course, her constant companion, y Môr, the Sea herself.

For the Sea keeps no secrets.

Stories ebb and flow with her tides.

On Calan Gaeaf.

Bibliography

Chwedl Dŵr/The Fairy Tale of Water

Condry, William, *The Natural History of Wales* (London, Collins New Naturalist Library, 1982)

Davies, Jonathan Ceredig, *Folk-lore of West and Mid-Wales* (Aberystwyth, Welsh Gazette, 1911)

Davies, Sioned, trans., *The Mabinogion* (Oxford University Press, 2007)

Dyfed Archaeological Trust, *Submerged Forests in Wales,* www.dyfedarchaeology. org.uk/lostlandscapes/submergedforests.html

Gower, Jon, *The Turning Tide: A Biography of the Irish Sea* (London, Harper Collins, 2023)

Imray, *Bristol Channel Chart Pack* (Imray, 2023 edition)

Imray, *North and West Wales Chart Pack* (Imray, 2023 edition)

Ings, Mike and Fran Murphy, *Tiroedd coll ein cyndadau, The Lost Lands of our Ancestors* (Llandeilo, Dyfed Archaeological Trust, 2011)

Jenkins, J. Geraint, *Nets and Coracles* (London, David & Charles, 1974)

Johnson, Donald S., *Phantom Islands of the Atlantic, The Legends of Seven Lands that Never Were* (Canada, Goose Lane, 1994)

Kavanagh, Erin, *Layers in the Landscape,* www.geomythkavanagh.com/ layers-in-the-landscape

Kavanagh, K.E. and Bates, M.R., 'Semantics of the Sea – Stories and Science along the Celtic Seaboard', *Internet Archaeology* 53, 2019, https://doi. org/10.11141/ia.53.8

Kay, Paul and Frances Dipper, *A Field Guide to the Marine Fishes of Wales and Adjacent Waters* (Llandybie, Goldstone Books, 2009)

Lillie, Malcolm, *Hunters, Fishers & Foragers in Wales* (Oxford, Oxbow, 2015)

Portalis, http://portalisproject.eu

Redknapp, Mark, Sian Rees and Alan Aberg, *Wales and the Sea, 10,000 Years of Welsh Maritime History* (Royal Commission on the Ancient and Historical Monuments of Wales, 2021)

Stevenson, Peter, *Welsh Folk Tales* (Stroud, History Press, 2017)

Stevenson, Peter I*llustrated Welsh Folk Tales for Young and Old* (Cheltenham, History Press, 2023)

Thomas, Gwyn and Margaret Jones, *Y Mabinogi* (Cyngor Celfyddydau Cymru, 1984)

Van Duzer, Chet, *Sea Monsters an Medieval and Renaissance Maps* (British Library, 2013)

Williams, Russ, *Where the Folk* (University of Wales, Calon Press, 2024)

1. The Mesolithic Mermaid and the Welsh Utopia

Benwell, Gwen and Arthur Waugh, *Sea Enchantress: The Tale of the Mermaid and her Kin* (London, Hutchinson, 1961)

Driver, Toby, *The Hillforts of Cardigan Bay* (Herefordshire, Logaston Press, 2016)

Jones, Gwyn, *Welsh Legends and Folk-Tales* (Oxford University Press, 1955), pp.216–222

Stevenson, Peter, *Ceredigion Folk Tales* (Stroud, History Press, 2014) pp.55–60, 69–71

Stevenson, *Welsh Folk Tales*, pp.29–45

Stevenson, Peter, *Y Dilyw/The Flood, Archaeoleg a chwedlau o Geredigion Fesolithig/ Archaeology and folk tales from Mesolithic Ceredigion* (Portalis, unpublished, 2023)

Rhys, Sir John, *Celtic Folklore: Welsh and Manx, Vol. 1* (Oxford, Henry Frowde, 1891), pp.151–168

2. Siani Chickens and the Ceredigion Storycatcher

Davies, Llinos M., *Crochan Ceredigion, Chwedlau i'r Hen a'r Ifanc* (Aberystwyth, Cyhoeddiadau Mei, 1992)

Evans, Myra, *Atgofion Ceinewydd* (Aberystwyth, Cambrian News, 1961), pp.32–37

Evans, Myra, *Casgliad o Chwedlau Newydd* (Aberystwyth, Cambrian News, 1926)

Evans, Myra, *Manuscript Collection*, various localities in private collections

James, David B., *Ceredigion – Its Natural History* (Llandre, James, 2001)

Leblond, Valériane and Peter Stevenson, *Shani Chickens* (Talybont, Y Lolfa, 2022)

Stevenson, *Ceredigion Folk Tales*, pp.70–71, 133–147, 163

Stevenson, *Y Dilyw/The Flood*, with special thanks to Portalis

3. The Submerged Land and the Wise Old Toad

Davies, Mared & Stevenson, Peter, *Y Gwraig i'r Eryr* (Llanrwst, Gwasg Carreg Gwalch, 2014)

Olding, Frank, *The Taliesin Sourcebook* (Somerset, Green Magic, 2014)

Stevenson, *Ceredigion Folk Tales*, pp.66–68, 78–82, 160

Thomas, Gwyn and Kevin Crossley-Holland, *The Tale of Taliesin* (London, Gollancz, 1992)

Thomas, Roger and Dudlyke, E.R., 'A Flint Chipping Floor at Aberystwyth' (*Journal of the Royal Anthropological Institute of Great Britain and Ireland*, Vol. 55, Jan–Jun, 1925, pp.73–89), www.jstor.org/stable/2843693

4. Migration Tales and the Cambrian Line

Holden, Chris, *The Essential Underwater Guide to North Wales, Barmouth to South Stack* (Calgo, 2003)

Hutton Catherine (1756–1846), *Letters Written During a Tour in North Wales, in Catherine Hutton's Tour of Wales*: 1796, NLW MS 19079C

Jarman, Eldra and Jarman, A.O.H., *The Welsh Gypsies* (Cardiff, University of Wales, 1991)

Joshi, Chandrika, *Ugandan Journeys* (People's Collection Wales, 2023), www.peoplescollection.wales/story/2007536

Roberts, Askew and Edward Woodall, *Gossiping Guide to Wales* (London, Simpkin, Marshall, Hamilton. Kent & Co., 1881)

Sampson, John, 'The Frozen Ship' (*Journal of the Gypsy Lore Society*, Vol. V11 No. 2, 1928)

5. A Train Journey to the Mabinogi

Booth, A.B. and Fowles, A.P., *A Review and Life History of the Rosy Marsh Moth, in The Moths of Ceredigion, Appendix 2* (Nature Conservancy Council, 1988)

Davies, *The Mabinogion*, pp.22–34

Griffiths, Gwyn, *The Last of the Onion Men* (Llanrwst, Gwasg Carreg Gwalch, 2002)

Maddern, Eric, *Snowdonia Folk Tales* (Stroud, History Press, 2015)

Redknapp et al, *Wales and the Sea*, pp.158–159, 170–171

Stevenson, *Welsh Folk Tales*, pp.122–123

Thomas and Jones, *Y Mabinogi*, pp.29–44

6. Island Tales of the Kings of Bardsey

Chamberlain, Brenda, *Tiderace* (London, Hodder & Stoughton, 1962)

Evans, Christine, *Bardsey* (Llandysul, Gomer, 2008)

Jones, Jennie, *Tomos o Enlli* (Llanrwst, Gwasg Carreg Gwalch, new edition 1999)

Jones, Peter Hope, *The Natural History of Bardsey* (Cardiff, National Museum of Wales, 1988)

Myrddin ap Dafydd, *Pobl Enlli* (Llanrwst, Gwasg Carreg Gwalch, 2015)

Stevenson, Peter, *Illustrated Welsh Folk Tales* (Cheltenham, History Press, 2023), pp.43–46, 225–227

Stevenson, Peter, *The Kings of Bardsey in Retracing Wales, Yr Eifl to Aberso*ch (Planet 208, 2012–13)

Stevenson, *Welsh Folk Tales*, pp.204–205

Stevenson, Illustrated Welsh Folk Tales, pp.43-46, 225-227

7. Rockpools and a Giants' Town on Pen Llŷn

Clowes, Carl, *Antur Aelhaearn* (Penygroes, Cyhoeddiadai Mei, 1982)
Clowes, Carl, *Nant Gwrtheyrn* (Talybont, Y Lolfa, 2004)
Evans, Hugh, *Y Tylwyth Teg* (Liverpool, Hugh Evans, 1935)
Rhys, Sir John, *Celtic Folklore: Welsh and Manx, Vol. 1* (Oxford, Henry Frowde, 1891), pp.176–196
Stevenson, Peter, *Boggarts, Trolls & Tylwyth Teg* (Stroud, History Press, 2021)
Stevenson, *Retracing Wales*, 2012–13)
Stevenson, *Welsh Folk Tales*, pp.67–69, 170–172, 238–239

8. Arianrhod and the Menai Strait Monsters

Davies, *The Mabinogion*, pp.47–64, 103–110
Emerson, P.H., *Welsh Fairy-Tales and other Stories* (London, D. Nutt, 1894)
Maddern, *Snowdonia Folk Tales* (Stroud, History Press, 2015)
Radford, Ken, *Tales of North Wales* (Letchworth, Garden City, 1982)
Stevens, Catrin, *Santes Dwynwen* (Llandysul, Gomer, 2005)
Stevenson, *Welsh Folk Tales*, pp.115–116
Thomas and Jones, *Y Mabinogi*, pp.64–85
Tunnicliffe, Charles, *Shorelands Summer Diary* (London, Collins, 1952)

9. Riding a Mammoth Round Ynys Môn

Constantine, Mary-Ann, *The Wreck of the Royal Charter, Ballads in Wales, Baledi yng Nghymru* (London: FLS Books, 1999, pp.65–85).
Jones, Alun R., *Lewis Morris* (Cardiff, University of Wales, 2004)
Jones, W. Eifion, *A New Natural History of Anglesey* (Anglesey Antiquarian Society, 1990)
Radford, *Tales of North Wales*, pp.81–86
Ren, https://rnli.org/news-and-media/2023/june/26/welsh-music-star-fundraiser-for-anglesey-rnli-crews-who-searched-for-best-friend
Stevenson, *Welsh Folk Tales*, p.56

10. Down the Rabbit Hole with a Goat

Holden, Chris, *The Essential Underwater Guide to North Wales, South Stack to Colwyn Bay* (Calgo, 2008)
Jarman, *The Welsh Gypsies*, 1991
Stevenson, Peter, Kavanagh, Erin, and Bates, Martin, *Retracing Wales: Llandudno* (Planet 218, Summer 2015, pp.43–49)
Senior, Michael, *Llys Helig and the Myth of Lost Lands* (Llanrwst, Gwasg Carreg Gwalch, 2002)
Stevenson, *Illustrated Welsh Folk Tales*, pp.25–28
Stevenson, *Welsh Folk Tales*, pp.157–159

11. The Dancing Girl and the Headless Hermit

Collins, Fiona, *Denbigh Folk Tales* (Stroud, History Press, 2011)
Curious Clwyd, www.mythslegendsodditiesnorth-east-wales.co.uk
Groome, Frances Hindes, *Gypsy Folk Tales* (London, Hurst & Blackett, 1899)

Pennant, Thomas, *A Tour in Wales* (H.D. Symonds, 1778–81)

Pennant, Thomas, *The History of the Parishes of Whiteford and Holywell* (Benjamin and J. White, 1796)

Stevenson, Peter, *The Moon-eyed People* (Stroud, History Press, 2019)

Stevenson, *Welsh Folk Tales*, pp.187–189

Rhys, Gruff, *American Interior* (London, Hamish Hamilton, 2014)

Rhys, *Celtic Folklore*, p.441

Williams, Gwyn A,. *Madoc: The Making of a Myth* (London, Methuen, 1980)

12. Sabrina and the Salmon Children

Black Rock Lave Net Fishermen, www.facebook.com/blackrocklavenets/?locale=en_GB

Jenkins, J. Geraint, *The Inshore Fishermen of Wales* (Cardiff, University of Wales, 1991)

Little, Cath, *Glamorgan Folk Tales* (Stroud, History Press, 1917)

Redknapp et al, *Wales and the Sea*, pp.120–121, 70–71, 76–77

Watkins, Christine, *Gwent Folk Tales* (Stroud, History Press, 2019)

Witts, Chris, *Sabrina*, www.severntales.co.uk/sabrina.html

Witts, Chris, *The Severn Estuary Crossings* (River Severn, 2017)

13. Temperance Town and Tiger Bay

Jones, Bethan, *Jessie Knight – The Lady Tattoo Artist, Amgueddfa Cymru 2023, https://museum.wales/blog/2564/Jessie-Knight---The-Lady-Tattoo-Artist*

Little, Cath, *Glamorgan Folk Tales* (Stroud, History Press, 1917)

Mohamud, Abdul, and Whitburn, Robin, *Doing Justice to History in Primary Schools (London, Trentham Books, 2016)*

Morais, Nia, *Betty: The Determined Life of Betty Campbell (Broga, 2023)*

Pankhurst, Richard and Ibrahim Ismaa'il., *An Early Somali Autobiography, in Africa 32, No. 2 (Centro Studi Paesi Extraeuropei, 1977)*

Stevenson, Peter, *Crankies in Wales, Henry Box Brown, www.thehistorypress.co.uk/articles/crankies-in-wales-the-man-who-brought-moving-panoramas-to-the-welsh-valleys*

Stevenson. *Boggarts, Trolls & Tylwyth Teg pp 176-8*

Stevenson, *Welsh Folk Tales, pp.119–120, 238–239*

St Fagans National Museum of History, https://museum.wales/stfagans

14. The Lady of Ogmore and the King of Porthcawl

Morgan, Alun, *Legends of Porthcawl and the Glamorgan Coast* (Cowbridge, D. Brown, 1974), pp.13–23, 35–39, 96–98

Price, Emily, Viva Porthcawl, *The Official Story of the Porthcawl Elvis Festival* (Harris, 2020)

Trevelyan, Marie, *Folk-lore and Folk-Stories of Wales* (London, Elliot Stock, 1909)

Radford, Ken, *Tales of South Wales* (London, Skilton and Shaw, 1979)

Stevenson, *Welsh Folk Tales*, pp.63–4, 161, 141, 223

15. Swansea Jack and Potato Jones

Ann of Swansea, *Cambrian Pictures or Every One Has Errors* (Dinas Powys, Honno, new edition 2021, first published 1810)

Hellier, Berni and Gayle Morgan Simmonds, *The True Tail of Swansea Jack* (Swansea Jack Publishing, 2022)

Joe's Ice Cream, www.joes-icecream.com

Kardomah café, www.kardomahcafe.com/about-us

Radford, *Tales of South Wales*, pp.5–7

Stevenson, *Boggarts Trolls & Tylwyth Teg*, pp.15–21

Stevenson, *Welsh Folk Tales*, pp.224–225, 240–241

Watkins, Graham, *Swansea and the Sea*, 10th July 2020, www.grahamwatkins.info/post/2015/07/23/swansea-and-the-sea

Watkins, Graham, *Swansea Jack*, 27th October 2021, www.grahamwatkins.info/post/2017/07/19/swansea-jack-1

16. Swan Girl and Water Horse of Gŵyr

Beynon, John, *From Bard to Verse* (Gower, John Beynon, 1996)

Jenkins, Nigel, *Real Gower* (Bridgend, Seren, 2014)

Mullard, Jonathan, *Gower* (London, Collins New Naturalist Library, 2006)

Radford, *Tales of South Wales*, pp.78–80, 83–84, 124–127, 148–149

Sommer, Marianne, *Bones and Ochre: The Curious Afterlife of the Red Lady of Paviland* (Harvard University Press, 2008)

Stevenson, *Welsh Folk Tales*, pp.70–71

Stevenson, *Illustrated Welsh Folk Tales*, pp.83–86

Watkins, Graham, *The Smugglers of Culver Hole*, 9th November 2018, www.grahamwatkins.info/post/2018/11/09/the-smugglers-of-culver-hole

17. Amelia and the Hell Dog of Laugharne

Earhart, Amelia, *20hrs. 40min. Our Flight in the Friendship* (New York, G.P. Putnam's Sons, 1928)

Jones, Edmund, *The Appearance of Evil, Apparitions of Spirits in Wales* (Cardiff, University of Wales, 2003)

Tucker, Ralph A, ed., *Laugharne Local History and Folk Lore* (Gomer, Llandysul, nd., written 1925)

Lewis, Siân, *Gwenllian – Warrior Princess* (Llanrwst, Gwasg Carreg Gwalch, 2011)

The Mock Mayor of Llansteffan, www.llansteffan.com/the-mock-mayor

Thomas, Dylan, *Ballad of the Long-Legged Bait, in Deaths and Entrances* (London, J.M. Dent, 1946)

18. Leekie Porridge and Betty Foggy

Allen, Richard C., *Nantucket Quakers and the Milford Haven Whaling Industry, 1791–1821* (Quaker Studies, Vol. 15 No. 1, 2010)

History Points, historypoints.org

James, David, *Pembroke Dockyard's Famous Ginkgo Tree* (People's Collection Wales, 2021), www.peoplescollection.wales/items/2004856

Jenkins, J. Geraint, *The Maritime Heritage of Dyfed* (Cardiff, National Museum of Wales, 1982)

John, Brian, *More Pembrokeshire Folk Tales* (Newport, Greencroft, 1996), pp.22, 43

John, Brian, *Pembrokeshire Folk Tales* (Newport, Greencroft, 1991), pp.120, 141–142

John, Brian, *The Last Dragon* (Newport, Greencroft, 1992), p.42, 104

Kightly, Charles, *A Mirror of Medieval Wales, Gerald of Wales and His Journey of 1188* (Cardiff, Cadw, 1988)

Narberth Museum, *Betty Foggy*, www.narberthmuseum.co.uk/betty-foggy

Phillips, Alan, *The Cinemas of West Wales* (Talybont, Y Lolfa, 2017)

Waverley, https://waverleyexcursions.co.uk

19. Skomer Oddy and the Smalls

Davies, *The Mabinogion*, p.34

Grooms, Chris, *The Giants of Wales, Cewri Cymru* (Cardiff, Edwin Mellen, 1993)

Hague, Douglas B., *Lighthouses of Wales, their architecture and archaeology* (Aberystwyth, Royal Commission, 1994)

John, *More Pembrokeshire Folk Tales*, p.74, 125

John, Brian, *Fireside Tales from Pembrokeshire* (Newport, Greencroft, 1993), pp. 8–9, 48–96, 138

John, *The Last Dragon*, p.25, 63

Jones, Gwyn, and Crossley-Holland, Kevin, *The Quest for Olwen* (Cambridge, Lutterworth, 1988), pp.50–59

Lockley, R.M., *Early Morning Island – or a dish of sprats* (London, Harrap, 1939)

Lockley, R.M., *Puffins* (London, J.M. Dent, 1953)

Lockley, R.M., *The Private Life of the Rabbit* (London, Andre Deutsch, 1964)

Mullard, Jonathan, *Pembrokeshire* (London, Collins New Naturalist Library, 2020)

Redknapp et al, *Wales and the Sea*, p.54

20. The Old Fibber and the Last Invasion

Carradice, Phil, *Britain's Last Invasion: The Battle of Fishguard 1797* (Barnsley, Pen & Sword, 2020)

John, *More Pembrokeshire Folk Tales*, pp.117–118

John, *Pembrokeshire Folk Tales*, pp.19–20, 53–54, 67–68, 78

Medlicott, Mary, *Shemi's Tall Tales* (Llandysul, Pont Books, 2008)

Redknapp et al, *Wales and the Sea*, pp.208–209

Stevenson, Peter, *Uisce Dŵr Water, Fibbing from Fishguard* (Aberystwyth, Centre for Welsh and Celtic Studies, 2022)

Stevenson, *Welsh Folk Tales*, p.31

Willison, Christine, *Pembrokeshire Folk Tales* (Stroud, History Press, 2013)

'Special thanks to Ports, Past and Present'

21. The Mermaid and the Swan Girl

Aaron, Jane, *Cranogwen* (Cardiff, University of Wales Press, 2023)

John, Brian, *Pembrokeshire Folk Tales* (Newport, Greencroft, 1992), pp.74–76

Matthias, Idris, *The Last of the Old Cardigan Ghosts* (Basye VA, Dolbadau Road Press, 2015)

Santschi-Cooney, Sascha, *The Forth and Bargy Dialect* (Wexford County Council, 2019)

Stevenson, *Ceredigion Folk Tales*, pp.61–65, 98–99

Stevenson, *Welsh Folk Tales*, pp.37–45

Stevenson, *Boggarts Trolls & Tylwyth Teg*, pp.207–212

Stevenson, *Uisce Dŵr Water*, pp.51–57

Thomas, David N., *Dylan Thomas: A Farm, two Mansions, and a Bungalow* (Bridgend: Seren, 2009)

More Websites

Beach buddies Wales, www.beachacademywales.com/buddies

British Sub Aqua Club Wales, www.bsac.com/this-is-bsac/bsac-team/regional-coaches/wales-region

Casgliad y Werin Cymru/People's Collection Wales, www.peoplescollection.wales

Coflein, https://coflein.gov.uk/en

Curious Travellers, curioustravellers.ac.uk

Dyfed Archeological Trust, https://dyfedarchaeology.org.uk

Early Tourists in Wales ,https://sublimewales.wordpress.com

Greenpeace Cardiff, www.greenpeace.org.uk

Heneb – The Trust for Welsh Archaeology, www.dyfedarchaeology.org.uk/wp

Marine Character Areas, https://naturalresources.wales/evidence-and-data/maps/marine-character-areas/?lang=en

Marine Conservation Society, www.mcsuk.org/about-us/where-we-work/our-focus-in-wales

North Wales Wildlife Trust Living Seas, www.northwaleswildlifetrust.org.uk/livingseas

National Trust Wales, www.nationaltrust.org.uk/visit/wales

National Waterfront Museum, https://museum.wales/swansea

Royal National Lifeboat Institution, rnli.org

Rights of Nature. Should the ocean have legal rights? The Revelator, 2002, https://therevelator.org/ocean-legal-rights

Royal Commission on the Ancient and Historical Monuments of Wales, https://rcahmw.gov.uk

Sea Trust Wales, www.seatrust.org.uk

Sea Watch Foundation, www.seawatchfoundation.org.uk

Welsh Coast Path, www.walescoastpath.gov.uk/?lang=en

Wildlife Trust of South and West Wales, www.welshwildlife.org/about-us/wildlife-conservation/living-seas

Wild Seas Wales, https://wildseas.wales/about

Thanks to the coastal storytellers: Carl Gough of Swansea, Cath Little of Cardiff, Chris and Macsen Baglin of Flint, Dafydd Davies Hughes of Aberdaron, Deb Winter of Fishguard, Gill Brownson of Holyhead, Phil Okwedy of Tenby.

Thanks also to Jacob Whittaker, Samantha Brummage, Martin Bates, and the Portalis archaeologists

Mesolithic Site Map

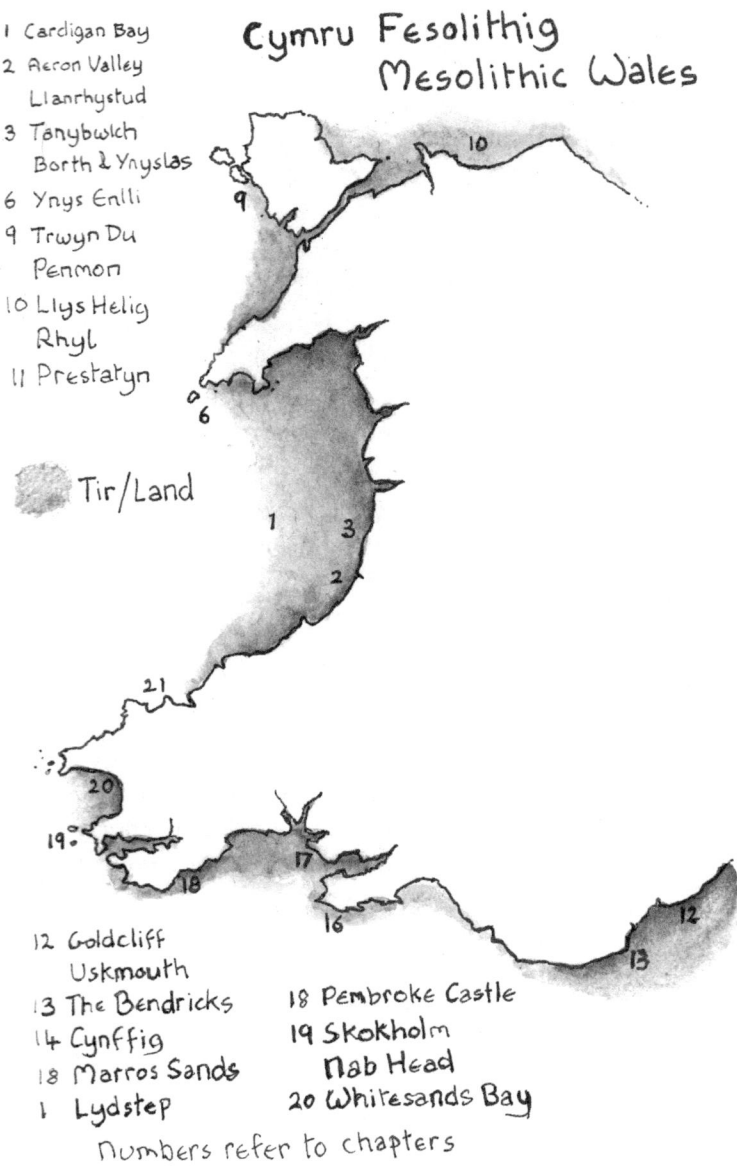

Cymru Fesolithig
Mesolithic Wales

1 Cardigan Bay
2 Aeron Valley
  Llanrhystud
3 Tanybwlch
  Borth & Ynyslas
6 Ynys Enlli
9 Trwyn Du
  Penmon
10 Llys Helig
  Rhyl
11 Prestatyn

Tir/Land

12 Goldcliff
   Uskmouth
13 The Bendricks
14 Cynffig
18 Marros Sands
1 Lydstep

18 Pembroke Castle
19 Skokholm
   Nab Head
20 Whitesands Bay

Numbers refer to chapters

Society *for* **Storytelling**

Since 1993, The Society for Storytelling has championed the ancient art of oral storytelling and its long and honourable history – not just as entertainment, but also in education, health, and inspiring and changing lives. Storytellers, enthusiasts and academics support and are supported by this registered charity to ensure the art is nurtured and developed throughout the UK.

Many activities of the Society are available to all, such as locating storytellers on the Society website, taking part in our annual National Storytelling Week at the start of every February, purchasing our quarterly magazine Storylines, or attending our Annual Gathering – a chance to revel in engaging performances, inspiring workshops, and the company of like-minded people.

You can also become a member of the Society to support the work we do. In return, you receive free access to Storylines, discounted tickets to the Annual Gathering and other storytelling events, the opportunity to join our mentorship scheme for new storytellers, and more. Among our great deals for members is a 30% discount off titles from The History Press.

For more information, including how to join, please visit

www.sfs.org.uk